GRAPHICAL MODELS FOR CATEGORICAL DATA

Alberto Roverato

SEMSTAT ELEMENTS

Ernst C. Wit

The Bernoulli Society

CAMBRIDGE
UNIVERSITY PRESS

Cambridge Elements ≡

CAMBRIDGE
UNIVERSITY PRESS

University Printing House, Cambridge CB2 8BS, United Kingdom

One Liberty Plaza, 20th Floor, New York, NY 10006, USA

477 Williamstown Road, Port Melbourne, VIC 3207, Australia

4843/24, 2nd Floor, Ansari Road, Daryaganj, Delhi – 110002, India

79 Anson Road, #06–04/06, Singapore 079906

Cambridge University Press is part of the University of Cambridge.

It furthers the University's mission by disseminating knowledge in the pursuit of education, learning, and research at the highest international levels of excellence.

www.cambridge.org
Information on this title: www.cambridge.org/9781108404969
DOI: 10.1017/9781108277495

First published 2017

A catalogue record for this publication is available from the British Library.

ISBN 978-1-108-40496-6 Paperback
ISSN 2514-3778 (Print)
ISSN 2398-404X (Online)

To Monica, Anna and Matteo

Contents

Preface

This book arises out of a short course given in a Séminaires Européens de Statistiques (SemStat) meeting at the European Institute for Statistics, Probability, Stochastic Operations Research and their Applications (EURANDOM) in Eindhoven, The Netherlands, over March 7–10, 2017. This SemStat meeting was organized as a part of the COST Action "European Cooperation for Statistics of Network Data Science" (COSTNET, CA15109) with the aim of introducing early career researchers to the field of statistical network science. In this perspective, the material presented here concerns the theory of graphical models and includes well-established methodology from the early developments in this field, but also the theory of models introduced more recently in the graphical model literature. The focus is on the discrete case where all the variables involved in the analysis are categorical and, in this context, classical and more recent results are presented in a unified way.

The approach followed in this text regarding the parameterization of the models is the result of a long-standing collaboration with Luca La Rocca and Monia Lupparelli on graphical models for discrete data. My views on this research area have also benefited from interactions with people of the "graphical model community," who shared with me their ideas and enthusiasm during these years. I would not be the author of this book were it not for the encouragement of Ernst Wit who also made valuable comments on the manuscript. Finally, I would like to thank my family, Monica, Anna and Matteo, for their support.

The material in this book is based upon work supported by the Air Force Office of Scientific Research under award number FA9550-17-1-0039.

Alberto Roverato
March 2017

1
Introduction

1.1 Graphical Models

Graphical models are an elegant framework that combines uncertainty and graph theory to represent complex phenomena. A graph is a structure consisting of a set of objects, called vertices, and a set of connections between pairs of vertices, called edges. The vertices of the graph associated with a graphical model are the variables of the model and the edges describe how the variables interact with each other. A further fundamental component of graphical models is the notion of independence or, more generally, of conditional independence. The edges missing from the graph can be interpreted as absence of interaction, in the sense that variables are conditionally independent. An appealing characteristic of graphs is that they can be represented graphically and many of their features and properties can be understood from the visual inspection of their graphical representation. The graph greatly simplifies the interpretation of the model, making its independence structure more immediate and intuitive. This also facilitates the communication of the scientific contents of the model to researchers who are not familiar with the statistical formalism. In large models the information provided by the visual inspection of the graph may be less clear but, nevertheless, the graph is useful in many ways. To mention a few, the graph structure may imply an intrinsic modularity that allows one to split the model into submodels of smaller dimensions, so that analyses of interest and statistical inference procedures can be carried out locally on marginal distributions. Furthermore, the graph is a natural object to be dealt with in the implementation of algorithms and computational procedures. All in all, graphical models constitute a very versatile methodology that has proved useful in a wide range of domains and applications. An historical overview of graphical models, with a comprehensive list of early references, can be found in Cox and

Wermuth (1996, section 2.13) and Lauritzen (1996, chapter 1), whereas we refer to Drton *et al.* (2017) for a recent collection of reviews.

1.2   Outline of the Book

Graphical modeling has been a very active area of research in the last years, and nowadays a wide range of families of graphical models are available; see Sadeghi and Lauritzen (2014). This book focuses on categorical data and describes the theory of some of the most relevant classes of graphical models. In this context, the book aims at providing a unified view of the theory discussed in a largely scattered literature.

The different families of models can be distinguished for the kind of graph associated with the probability distribution. The models considered in Chapters 4 and 5 deal with symmetric relationships between variables represented by undirected and bidirected graphs, respectively. Undirected graph models, also known as Markov random fields, are characterized by collections of conditional independence relationships, whereas bidirected graph models are characterized by collections of marginal independencies. Asymmetric relationships between variables are introduced in Chapter 6. Firstly, we consider the family of models associated with directed acyclic graphs (DAGs), also known as Bayesian networks, and then models associated with regression graphs. The latter family includes each of the previous families of models as special case and, therefore, regression graph models constitute a general framework that unifies and extends the theory of undirected, bidirected and directed acyclic graph models.

1.2.1   Discrete Graphical Models and Their Parameterization

The graph associated with the model allows one to abstract the conditional independence relationships between variables from the details of their parametric forms. Accordingly, the interpretation of a graphical model and the inferential questions of interest usually do not depend on the kinds of variables involved in the analysis. On the other hand, the effective specification of the statistical model and the implementation of statistical techniques require the definition of a suitable parameterization that cannot prescind from the variable types. From this perspective, the attention is restricted to categorical variables, that are variables taking one of a finite number of possible values. Our approach to the specification of parameterizations aims at building a common framework encompassing all the statistical models considered. Chapter 2 deals with the probability distribution of a random vector in the form of a probability table, and provides a set of rules for establishing conditional independence, based on cross-product ratios. Chapter 3 gives the theory of Möbius inversion that is then used to transform probabilities into more convenient log-linear parameterizations. Jointly, Chapters 2 and 3 provide a set of tools that can be applied, almost identically, to all the families of models we consider to obtain the corresponding parameterizations and derive their properties.

1.2.2 *Binary vs Non-binary Variables*

In its simplest from, a categorical variable takes one of two possible values and is thus called a binary variable. Dealing with binary variables is especially convenient because, in this setting, the notation is less involved and the presentation of the material is more immediate. Furthermore, it is often the case that the properties of the binary case can be used as building blocks for the derivation of the corresponding properties for the general case. Throughout this text, we keep a clear distinction between the general case of arbitrary categorical variables and the special case of binary variables. We deem that this approach may facilitate the comprehension of the material because the treatment of the general case can often be regarded as a mere technical generalization of the binary case. For this reason, when suitable, we will first discuss the binary case in detail, and then we will consider the general case in a separate subsection, which the reader may choose to skip on first reading.

2

Conditional Independence and Cross-product Ratios

In this chapter we introduce the notation and the terminology used to deal with categorical variables, and give the notions of independence and conditional independence, which play a central role in the theory of graphical models.

One important issue concerning the characterization of families of graphical models is the specification of suitable parameterizations that may allow one to efficiently deal with the collection of independence relationships defining the model. Although different families of graphical models require different parameterizations, for the models we consider it is possible to construct a unified approach based on the connection between independence relationships and cross-product ratios, which can be implemented by means of a tool called Möbius inversion. One advantage of this framework is that it leads to a simplification of the theory, in the sense that several fundamental results follow immediately from the application of a few basic lemmas connecting cross-product ratios to Möbius inversion. On the other hand, this approach sheds light on some common features of different families of graphical models, which are not usually presented jointly in the literature.

The material concerning cross-product ratios is given here, whereas the theory of Möbius inversion and the derivation of its relevant properties are deferred to the next chapter.

2.1 Notation and Terminology

This section is devoted to illustrating the notation we use to deal with vectors and cross-classified tables. It is important that the readers feel comfortable with these conventions. In our notation, vectors and matrices are indexed by the elements of a finite set. For instance, for a finite set V with cardinality $p = |V|$ we write $Y_V = (Y_v)_{v \in V}$ to denote a vector of p

random variables indexed by the elements of V, and we will simply say that Y_V *is indexed by* V. Furthermore, we denote by 0_V the vector of length p such that all entries are equal to zero and indexed by an element of V and, similarly, 1_V is a vector with all entries equal to one. Any subset $A \subseteq V$ with $A \neq \emptyset$ identifies the subvector $Y_A = (Y_v)_{v \in A}$ of Y_V; we regard vectors as column vectors and denote by Y_A^T the transpose of Y_A.

2.1.1 Cross-classified Tables

Let $Y_V = (Y_v)_{v \in V}$ be a vector of categorical random variables indexed by the finite set V with $p = |V|$. Categorical variables can take a finite number of values that are usually called *levels*. For notational convenience, we label the levels of the variables Y_v, for $v \in V$, as $0, 1, \ldots, d_v$ so that $\mathcal{I}_v = \{0, 1, \ldots, d_v\}$ is the *state space* of Y_v and the Cartesian product $\mathcal{I}_V = \times_{v \in V} \mathcal{I}_v$ is the state space of Y_V. We will refer to $\mathcal{I}_V$ as a p-dimensional *cross-classified table*, or simply *table* for short. Accordingly, the elements of $\mathcal{I}_V$, denoted by $i_V = (i_v)_{v \in V}$, are called the *cells* of the table. The state space of the subvector Y_A, with $A \subseteq V$ and $A \neq \emptyset$, is the *marginal table* $\mathcal{I}_A = \times_{v \in A} \mathcal{I}_v$ with cells $i_A \in \mathcal{I}_A$.

A random variable is called *binary* or *dichotomous* if it takes only two possible values and we say that a variable is *non-binary* or *polytomous* when it can take an arbitrary, finite, number of values. We will start by considering vectors of binary variables, and use the subsets of V to index the cells of the associated tables. Then, we will consider the general case of non-binary vectors and introduce a notation that will allow us to represent the associated table as a collection of binary subtables. This approach facilitates both the presentation of the material and the proof of the results, typically given for the special case of binary variables, and then naturally extended to the general polytomous case by repeated application of the result for the binary case to all the binary subtables.

The Binary Case

Every binary variable takes values in the set $\{0, 1\}$ and the state space of Y_V is the Cartesian product $\{0, 1\}^p$, which we call a p-dimensional *binary table*. A fundamental feature of the notation we use is that the entries of Y_V are indexed by the elements of V so that the cells of the binary table $\mathcal{I}_V$ associated with Y_V can be indexed by subsets of V; i.e., by the power set of V. Concretely, we use the convention that the cell associated with the subset $D \subseteq V$ is the vector indexed by V with all the entries indexed by D equal to one, and all the entries indexed by $V \setminus D$ equal to zero. In this way, every subset $D \subseteq V$ uniquely identifies a cell $(1_D, 0_{V \setminus D})$, and we can write the binary table associated with Y_V as

$$\mathcal{I}_V = \{(1_D, 0_{V \setminus D})\}_{D \subseteq V}. \tag{2.1}$$

Notice that vectors indexed by the empty set are void and, therefore, when $D = \emptyset$ it holds that $(1_D, 0_{V \setminus D}) = 0_V$, whereas if $D = V$ then $(1_D, 0_{V \setminus D}) = 1_V$. An instance of three-way binary table can be found in Table 2.1. In this example, every cell of the table contains the corresponding indexing set, and one should notice that the indexing sets identify the variables whose levels are equal to one, whereas all the variables not included in the indexing set take value zero.

In the remainder of this text we will use some simplifications in order to lighten the notation. First, unless we deem it useful to highlight it, we will omit the subscript when it is equal to V so that, for instance, $Y = Y_V$, $\mathcal{I} = \mathcal{I}_V$ and $i = i_V$. Second, for a subset $D \subseteq V$ we will write $Y_D = 1$ to denote $Y_D = 1_D$, that is, that all the entries of Y_D take value one. Similarly, we will write $Y_D = 0$ in place of $Y_D = 0_D$. Finally, we will omit the brackets from singleton sets, as already done in Table 2.1.

The Non-binary Case

The case of categorical variables with arbitrary number of levels can be conveniently dealt with by introducing a version of the sample space deprived of one level for every variable. Formally, the *restricted state space* of Y_v is defined as $\mathcal{J}_v = \mathcal{I}_v \setminus \{0\} = \{1, \ldots, d_v\}$ so that the restricted state space of Y is given by $\mathcal{J} = \mathcal{J}_V = \times_{v \in V} \mathcal{J}_v$. Hereafter,

Table 2.1 Indexing sets of the cells of the cross-classified table associated with the binary vector $Y_{\{a,b,c\}}$.

	$Y_c = 0$			$Y_c = 1$	
	Y_b			Y_b	
Y_a	0	1	Y_a	0	1
0	$\emptyset$	b	0	c	$\{b, c\}$
1	a	$\{a, b\}$	1	$\{a, c\}$	$\{a, b, c\}$

we will refer to "0" as the *baseline* level of Y_v and we remark that the choice of the level to be set as baseline is arbitrary. The elements of $\mathcal{J}$ are denoted by $j = j_V$ and for every $A \subseteq V$, with $A \neq \emptyset$, we denote by $\mathcal{J}_A$ the restricted state space of Y_A. Every pair $j \in \mathcal{J}$ and $D \subseteq V$ uniquely identifies the cell $(j_D, 0_{V \setminus D}) \in \mathcal{I}$, with the usual convention that for $D = \emptyset$ the vector j_D is void, and therefore $(j_\emptyset, 0_V) = 0_V$. Hence, every $j \in \mathcal{J}$ uniquely identifies the subtable

$$\{(j_D, 0_{V \setminus D})\}_{D \subseteq V}, \tag{2.2}$$

and with a slight abuse of terminology we will refer to (2.2) as a *binary subtable*. Table 2.2 gives an instance of the binary subtable identified by $j_V \in \mathcal{J}_V$ with $V = \{a, b, c\}$. It follows that the introduction of the restricted state space allows us to write the state space of Y_V as a collection of binary subtables, one for every $j \in \mathcal{J}$; formally,

$$\mathcal{I} = \{(j_D, 0_{V \setminus D})\}_{j \in \mathcal{J}, D \subseteq V}. \tag{2.3}$$

Clearly, in the case of binary variables, (2.3) coincides with (2.1) because the restricted state space $\mathcal{J}$ has only one element, that is the vector 1_V.

The representation of $\mathcal{I}$ as the collection of the binary subtables $\{(j_D, 0_{V \setminus D})\}_{D \subseteq V}$, for $j \in \mathcal{J}$, is useful in practice because we will see that several relevant properties that hold in the binary case can be immediately generalized to the non-binary case by their iterative application for $j \in \mathcal{J}$. On the other hand, we remark that this representation of $\mathcal{I}$ is not efficient because some cells are simultaneously present in more than one binary subtable. The most relevant example of this situation is given by

Table 2.2 Indexes of the cells of the subtable identified by $j_V \in \mathcal{J}_V$ with $V = \{a, b, c\}$.

	$Y_c = 0$			$Y_c = j_c$	
		Y_b			Y_b
Y_a	0	j_b	Y_a	0	j_b
0	$0_{\{a,b,c\}}$	$(j_b, 0_{\{a,c\}})$	0	$(j_c, 0_{\{a,b\}})$	$(j_{\{b,c\}}, 0_a)$
j_a	$(j_a, 0_{\{b,c\}})$	$(j_{\{a,b\}}, 0_c)$	j_a	$(j_{\{a,c\}}, 0_b)$	$j_{\{a,b,c\}}$

the cell 0_V that is contained in every subtable $\{(j_D, 0_{V \setminus D})\}_{D \subseteq V}$ for $j \in \mathcal{J}$.

2.2 Conditional Independence

We introduce now the important notions of independence and conditional independence of random variables. Then, we characterize a family of cross-product ratios and show how these quantities can be used to assess independence relationships.

The probability mass function of a categorical random vector Y_V can be given in the form of a *probability table*

$$\{p(Y = i)\}_{i \in \mathcal{I}}$$

and, throughout this text, we restrict our attention to random vectors with positive probability mass function, that is

$$p(Y = i) > 0 \quad \text{for all } i \in \mathcal{I}.$$

The *marginal probability table* of the subvector Y_A is $\{p(Y_A = i_A)\}_{i_A \in \mathcal{I}_A}$, and if A and B are two disjoint subsets of V, that is $A, B \subseteq V$ with $A \cap B = \emptyset$, then the subvector $Y_{A \cup B}$ can also be written as (Y_A, Y_B) and, therefore, the forms $p(Y_{A \cup B} = i_{A \cup B})$ and $p(Y_A = i_A, Y_B = i_B)$ are equivalent. Accordingly, if $B = \emptyset$ then $p(Y_A = i_A, Y_B = i_B) = p(Y_A = i_A)$. More generally, for every $B \subseteq V \setminus A$,

$$p(Y_A = i_A) = \sum_{i_B \in \mathcal{I}_B} p(Y_A = i_A, Y_B = i_B),$$

and thus it makes sense to use the convention that $p(Y_A = i_A) = 1$ when $A = \emptyset$.

The notions of independence and conditional independence of random variables are central to the interpretation of graphical models.

Definition 2.2.1 Let Y_V be a vector of categorical random variables and let A and B be two nonempty, disjoint, subsets of V. We say that Y_A *is independent of* Y_B if and only if

$$p(Y_A = i_A, Y_B = i_B) = p(Y_A = i_A)p(Y_B = i_B), \tag{2.4}$$

for every $i_A \in \mathcal{I}_A$ and $i_B \in \mathcal{I}_B$.

It is straightforward to see that equation (2.4) is equivalent to either of the equations

$$p(Y_A = i_A | Y_B = i_B) = p(Y_A = i_A) \tag{2.5}$$

and

$$p(Y_B = i_B | Y_A = i_A) = p(Y_B = i_B),$$

each of which can thus replace (2.4) to obtain two alternative, commonly used, ways to define the independence of random vectors. Hereafter, we shall use the notation introduced by Dawid (1979), and write $Y_A \perp\!\!\!\perp Y_B$ to denote the independence of Y_A and Y_B.

The notion of independence can be generalized to that of conditional independence, that is the independence of two random vectors Y_A and Y_B for each fixed value of a third vector Y_C.

Definition 2.2.2 Let Y_V be a vector or categorical random variables, and let A, B and C be three nonempty, pairwise disjoint, subsets of V. We say that Y_A is *conditionally independent of Y_B given Y_C*, and write $Y_A \perp\!\!\!\perp Y_B | Y_C$, if and only if

$$\begin{aligned} p(Y_A = i_A, Y_B = i_B | Y_C = i_C) \\ = p(Y_A = i_A | Y_C = i_C) p(Y_B = i_B | Y_C = i_C), \end{aligned} \tag{2.6}$$

for every $i_A \in \mathcal{I}_A$, $i_B \in \mathcal{I}_B$ and $i_C \in \mathcal{I}_C$.

As well as for the definition of independence, alternative ways to define conditional independence are obtained by replacing equation (2.6) by either

$$p(Y_A = i_A | Y_B = i_B, Y_C = i_C) = p(Y_A = i_A | Y_C = i_C),$$

or

$$p(Y_B = i_B | Y_A = i_A, Y_C = i_C) = p(Y_B = i_B | Y_C = i_C).$$

2.3 Establishing Independence Relationships

We now give some alternative ways to establish independence relationships that will lead to the introduction of a relevant class of cross-product ratios. We start from the binary case and then extend the results to the general case of polytomous variables.

Table 2.3 Two-by-two probability table of $Y_{\{a,b\}}$.

		Y_b	
Y_a	0	1	
0	$p(Y_a = 0, Y_b = 0)^{*}$	$p(Y_a = 0, Y_b = 1)^{*\ddagger}$	$p(Y_a = 0)^{\ddagger}$
1	$p(Y_a = 1, Y_b = 0)^{*\S}$	$p(Y_a = 1, Y_b = 1)^{*\dagger\ddagger\S}$	$p(Y_a = 1)^{\dagger\ddagger}$
	$p(Y_b = 0)^{\S}$	$p(Y_b = 1)^{\S\dagger}$	$1^{\dagger}$

The Binary Case

The cells of the cross-classified table associated with a binary vector Y_V are indexed by the subsets of V and, accordingly, the probability table can be written as

$$\{p(Y_D = 1, Y_{V \setminus D} = 0)\}_{D \subseteq V}.$$

There exist several alternative ways to establish the independence of two binary random vectors. For example, consider the probabilities for the binary variables Y_a and Y_b given in Table 2.3. It is easy to check that a necessary and sufficient condition for the independence of Y_a and Y_b is that the following identity is satisfied

$$p(Y_a = 1 | Y_b = 1) = p(Y_a = 1 | Y_b = 0). \tag{2.7}$$

Another equality that constitutes a necessary and sufficient condition for the independence of these two variables is

$$p(Y_a = 1 | Y_b = 1) = p(Y_a = 1). \tag{2.8}$$

From our viewpoint, the relevant difference between equations (2.7) and (2.8) is that the former only involves the conditional distributions of $Y_a | Y_b$, whereas the latter also involves a marginal distribution of Y_a. We can, rather informally, say that equation (2.7) originates from a "conditional approach," whereas (2.8) comes from a "marginal approach," and this distinction becomes especially relevant in multivariate settings with more than two variables. The probability distribution of a categorical random vector can be given in the form of a collection of marginal and conditional probabilities. This representation is not unique and the collection of probabilities characterizing the distribution may be suitably chosen so as to facilitate dealing with the type of independence relationships of interest, which may be marginal independencies between subsets of variables, conditional independencies involving all the variables or a combination of these two situations. This issue can be approached with the proper level of generality by introducing the two functions $c(\cdot)$ and $m(\cdot)$ defined on the power set of V. More precisely, for every $D \subseteq V$, the function $c(\cdot)$ is such that $c(D) = D$, whereas $m(\cdot)$ is such that $m(D) = \emptyset$. Then, if A is a subset of V and $A' \subseteq A$ we define

$$p_f(A',A) = p(Y_{A'} = 1, Y_{f(A\setminus A')} = 0),$$

where $f(\cdot)$ can be either $c(\cdot)$ or $m(\cdot)$, which we compactly write as $f \in \{c,m\}$. Notice that

$$p_m(A',A) = p(Y_{A'} = 1)$$

is a probability from the marginal distribution of the subvector $Y_{A'}$ of Y_A, whereas

$$p_c(A',A) = p(Y_{A'} = 1, Y_{A\setminus A'} = 0)$$

is a probability from the joint distribution of Y_A. Furthermore, if B is a subset of V such that $A \cap B = \emptyset$, and $B' \subseteq B$ then we set

$$p_{fg}(A',A : B',B) = p(Y_{A'} = 1, Y_{f(A\setminus A')} = 0, Y_{B'} = 1, Y_{g(B\setminus B')} = 0),$$

$$(2.9)$$

so that the pair $f,g \in \{c,m\}$ can be suitably chosen to specify four different probabilities. Specifically, $p_{cc}(A',A : B',B)$ is a probability from the joint distribution of both Y_A and Y_B, $p_{cm}(A',A : B',B)$ is a probability from the joint distribution of Y_A and the marginal distribution of $Y_{B'}$, $p_{mc}(A',A : B',B)$ is a probability from the marginal distribution of $Y_{A'}$ and the joint distribution of Y_B and, finally, $p_{mm}(A',A : B',B)$ is a probability from the marginal distributions of $Y_{A'}$ and $Y_{B'}$. Note also that for $B' = \emptyset$ and $g = m$ equation (2.9) simplifies to

$$p_{fm}(A',A : \emptyset,B) = p_f(A',A) \text{ for every } f \in \{c,m\},$$

whereas $p_{mg}(\emptyset,A : B',B) = p_g(B',B)$, with the usual convention that $p_{mm}(\emptyset,A : \emptyset,B) = 1$. From (2.9) we can define the conditional probability

$$p_{fg}(A',A|B',B) = p(Y_{A'} = 1, Y_{f(A\setminus A')} = 0|Y_{B'} = 1, Y_{g(B\setminus B')} = 0).$$

$$(2.10)$$

More generally, for $f, g \in \{c, m\}$, equation (2.10) takes four distinct forms, specifically,

$$p_{cc}(A',A|B',B) = p(Y_{A'} = 1, Y_{A\setminus A'} = 0|Y_{B'} = 1, Y_{B\setminus B'} = 0)$$

$$p_{mc}(A',A|B',B) = p(Y_{A'} = 1|Y_{B'} = 1, Y_{B\setminus B'} = 0)$$

$$p_{cm}(A',A|B',B) = p(Y_{A'} = 1, Y_{A\setminus A'} = 0|Y_{B'} = 1)$$

$$p_{mm}(A',A|B',B) = p(Y_{A'} = 1|Y_{B'} = 1).$$

Hence, the equalities (2.7) and (2.8) can be expressed in terms of (2.10) because (2.7) coincides with both

$$p_{cc}(a, a|b, b) = p_{cc}(a, a|\varnothing, b) \text{ and } p_{mc}(a, a|b, b) = p_{mc}(a, a|\varnothing, b),$$

whereas (2.8) coincides with both

$$p_{cm}(a, a|b, b) = p_{cm}(a, a|\varnothing, b)$$

and

$$p_{mm}(a, a|b, b) = p_{mm}(a, a|\varnothing, b).$$

The following lemma provides a general rule for establishing the independence of two random vectors through (2.10). This includes both (2.7) and (2.8) as special cases.

Lemma 2.1 *Let Y_V be a random vector of binary variables and let A and B be two nonempty, disjoint, subsets of V. Then, for any pair $f, g \in \{c, m\}$ it holds that $Y_A \perp\!\!\!\perp Y_B$ if and only if*

$$p_{fg}(A',A|B',B) = p_{fg}(A',A|\varnothing, B),$$

for every $A' \subseteq A$ and $B' \subseteq B$ such that $A', B' \neq \varnothing$.

Proof. We first show that $Y_A \perp\!\!\!\perp Y_B$ implies the case $(f, g) = (m, m)$. More concretely, we have to show that $Y_A \perp\!\!\!\perp Y_B$ implies $p(Y_{A'} = 1|Y_{B'} = 1) = p(Y_{A'} = 1)$ for every $\varnothing \neq A' \subseteq A$ and $\varnothing \neq B' \subseteq B$ and this follows from (2.5) because $Y_A \perp\!\!\!\perp Y_B$ implies $Y_{A'} \perp\!\!\!\perp Y_{B'}$.

Second, we show that the case $(f, g) = (m, m)$ is equivalent to the case $(f, g) = (c, m)$. The proof that the case $(f, g) = (c, m)$ implies the case $(f,g) = (m,m)$ is straightforward because it follows by taking the sum over the subsets $A'' \subseteq A \backslash A'$ of both sides of the equality

$$p(Y_{A'} = 1, Y_{A''} = 1, Y_{A \backslash (A' \cup A'')} = 0 | Y_{B'} = 1)$$
$$= p(Y_{A'} = 1, Y_{A''} = 1, Y_{A \backslash (A' \cup A'')} = 0)$$

that, in the case $(f,g) = (c,m)$, holds for every $\emptyset \neq A' \subseteq A$ and $\emptyset \neq B' \subseteq B$. We now show the reverse implication and, more concretely, that

$$p(Y_{A'} = 1 | Y_{B'} = 1) = p(Y_{A'} = 1) \tag{2.11}$$

for every $\emptyset \neq A' \subseteq A$ and $\emptyset \neq B' \subseteq B$ implies $p(Y_{A'} = 1, Y_{A \backslash A'} = 0 | Y_{B'} = 1) = p(Y_{A'} = 1, Y_{A \backslash A'} = 0)$ for every $\emptyset \neq A' \subseteq A$ and $\emptyset \neq B' \subseteq B$. The proof of this point is by induction on the cardinality of $A \backslash A'$ which, in the rest of this proof, we will denote by $\overline{A} = A \backslash A'$. The result is trivially true for $A' = A$, that is for $|\overline{A}| = 0$, and we show that if the result is true for $|\overline{A}| = k$ with $k \geq 0$ then it is true for $|\overline{A}| = k + 1$. If $|\overline{A}| = k + 1$ then $\overline{A} \neq \emptyset$ and we can choose an arbitrary element $a \in \overline{A}$ to write

$$p(Y_{A'} = 1, Y_{\overline{A}} = 0 | Y_{B'} = 1)$$
$$= p(Y_{A'} = 1, Y_{\overline{A} \backslash \{a\}} = 0, Y_a = 0 | Y_{B'} = 1)$$
$$= p(Y_{A'} = 1, Y_{\overline{A} \backslash \{a\}} = 0 | Y_{B'} = 1)$$
$$- p(Y_{A'} = 1, Y_{\overline{A} \backslash \{a\}} = 0, Y_a = 1 | Y_{B'} = 1)$$
$$= p(Y_{A'} = 1, Y_{\overline{A} \backslash \{a\}} = 0) - p(Y_{A'} = 1, Y_{\overline{A} \backslash \{a\}} = 0, Y_a = 1) \tag{2.12}$$
$$= p(Y_{A'} = 1, Y_{\overline{A}} = 0)$$

where (2.12) follows from the induction assumption because $|\overline{A} \backslash \{a\}| = k$.

Third, we show that the case $(f,g) = (c,m)$ implies the case $(f,g) = (m,c)$. It is easy to see that if the equality (2.11) holds true, then the same equality also holds if the subsets A and B are swapped. Because the case $(f,g) = (m,m)$ is equivalent to the case $(f,g) = (c,m)$, it follows that we can also swap the subsets A and B in the latter case. Hence, we have to show that

$$p(Y_{B'} = 1, Y_{B \backslash B'} = 0 | Y_{A'} = 1) = p(Y_{B'} = 1, Y_{B \backslash B'} = 0) \tag{2.13}$$

for every $\emptyset \neq A' \subseteq A$ and $\emptyset \neq B' \subseteq B$ implies $p(Y_{A'} = 1|Y_{B'} = 1,$ $Y_{B\backslash B'} = 0) = p(Y_{A'} = 1|Y_B = 0)$ for every $\emptyset \neq A' \subseteq A$ and $\emptyset \neq B'$ $\subseteq B$. We first notice that the identity (2.13) also holds for $B' = \emptyset$. This can be seen by summing both sides of (2.13) over $B' \subseteq B$, with $B' \neq \emptyset$, to obtain $1 - p(Y_B = 0|Y_{A'} = 1) = 1 - p(Y_B = 0)$ and therefore $p(Y_B = 0|Y_{A'} = 1) = p(Y_B = 0)$. Hence, we can multiply both sides of (2.13) by $p(Y_{A'} = 1)$ and divide by $p(Y_{B'} = 1, Y_{B\backslash B'} = 0)$ to obtain

$$p(Y_{A'} = 1|Y_{B'} = 1, Y_{B\backslash B'} = 0) = p(Y_{A'} = 1) \qquad (2.14)$$

for every $B' \subseteq B$ and $\emptyset \neq A' \subseteq A$. For $B' = \emptyset$ the equality (2.14) becomes $p(Y_{A'} = 1|Y_B = 0) = p(Y_{A'} = 1)$ and therefore we can replace $p(Y_{A'} = 1)$ by $p(Y_{A'} = 1|Y_B = 0)$ in (2.14) to obtain that

$$p(Y_{A'} = 1|Y_{B'} = 1, Y_{B\backslash B'} = 0) = p(Y_{A'} = 1|Y_B = 0) \qquad (2.15)$$

for every $A' \subseteq A$ and $\emptyset \neq B' \subseteq B$, as required.

In order to show that the case $(f,g) = (m,c)$ implies the case $(f,g) = (c,c)$, it is sufficient to use the same induction argument used to show that the case $(f,g) = (m,m)$ implies the case $(f,g) = (c,m)$.

Finally, we show that the case $(f,g) = (c,c)$ implies $Y_A \perp\!\!\!\perp Y_B$. We first notice that the equality

$$p(Y_{A'} = 1, Y_{A\backslash A'} = 0|Y_{B'} = 1, Y_{B\backslash B'} = 0)$$
$$= p(Y_{A'} = 1, Y_{A\backslash A'} = 0|Y_B = 0) \qquad (2.16)$$

holds true for every $A' \subseteq A$ and $B' \subseteq B$. Indeed, under the case $(f,g) = (c,c)$ the identity (2.16) is true for every $\emptyset \neq A' \subseteq A$ and $\emptyset \neq B' \subseteq B$. Furthermore, (2.16) is trivially true in the case $B' = \emptyset$ and, finally, it can be shown that it holds true also when $A' = \emptyset$ by using the same argument used to show that the identity (2.13) holds true for $A' = \emptyset$.

Then, one can compute $p(Y_{A'} = 1, Y_{A\backslash A'} = 0)$ by applying the low of total probability to show that this probability is equal to $p(Y_{A'} = 1, Y_{A\backslash A'} = 0|Y_B = 0)$, thereby obtaining

$$p(Y_{A'} = 1, Y_{A\backslash A'} = 0|Y_{B'} = 1, Y_{B\backslash B'} = 0) = p(Y_{A'} = 1, Y_{A\backslash A'} = 0),$$

for every $A' \subseteq A$ and $B' \subseteq B$, and this completes the proof. $\qquad \square$

A fundamental measure of the association of two binary variables Y_a and Y_b is the *odds ratio* that is the ratio of the products

$p(Y_a = 0, Y_b = 0)p(Y_a = 1, Y_b = 1)$ and $p(Y_a = 0, Y_b = 1)p(Y_a = 1, Y_b = 0)$. It is well-known that $Y_a \perp\!\!\!\perp Y_b$ if and only if the odds ratio is equal to one, that is

$$\frac{p(Y_a = 0, Y_b = 0)p(Y_a = 1, Y_b = 1)}{p(Y_a = 0, Y_b = 1)p(Y_a = 1, Y_b = 0)} = 1; \qquad (2.17)$$

see Agresti (2013) for a comprehensive treatment of this issue. The odds ratio is also called the *cross-product ratio* because it is the ratio of the products of probabilities from diagonally opposite cells, which are marked by an " $*$ " in Table 2.3. The odds ratio is the ratio of the cross-products of the joint probabilities in Table 2.3. We introduce three alternative cross-product ratios involving also the marginal probabilities of the two variables. Each of these cross-product ratios leads to an equality that is equivalent to the independence of the two variables.

We first consider the cross-product ratio of the probabilityes marked by a " ‡ " in Table 2.3 which involves the marginal distribution of Y_a, but not that of Y_b, and it holds that $Y_a \perp\!\!\!\perp Y_b$ if and only if

$$\frac{p(Y_a = 0)p(Y_a = 1, Y_b = 1)}{p(Y_a = 0, Y_b = 1)p(Y_a = 1)} = 1. \qquad (2.18)$$

Second, we consider the cross-product ratio of the probabilities marked by a "§". This involves the marginal distribution of Y_b, but not that of Y_a, and it holds that $Y_a \perp\!\!\!\perp Y_b$ if and only if

$$\frac{p(Y_b = 0)p(Y_a = 1, Y_b = 1)}{p(Y_b = 1)p(Y_a = 1, Y_b = 0)} = 1. \qquad (2.19)$$

Third, we consider the cross-product ratio of the probabilities marked by a "†". This involves the marginal distributions of both Y_a and Y_b and, also in this case, it holds that $Y_a \perp\!\!\!\perp Y_b$ if and only if

$$\frac{1 \times p(Y_a = 1, Y_b = 1)}{p(Y_b = 1)p(Y_a = 1)} = 1. \qquad (2.20)$$

It is possible to give a unified, and more general, version of the cross-product ratios above as follows

$$\mathrm{cpr}_{fg}(A', A : B', B) = \frac{p_{fg}(\varnothing, A : \varnothing, B)\, p_{fg}(A', A : B', B)}{p_{fg}(\varnothing, A : B', B)\, p_{fg}(A', A : \varnothing, B)} \qquad (2.21)$$

where $f, g \in \{c, m\}$. Hence, the equality (2.17) can be written as $\mathrm{cpr}_{cc}(a, a : b, b) = 1$ and, furthermore, the equalities (2.18), (2.19) and (2.20) can be written as $\mathrm{cpr}_{cm}(a, a : b, b) = 1$, $\mathrm{cpr}_{mc}(a, a : b, b) = 1$ and $\mathrm{cpr}_{mm}(a, a : b, b) = 1$, respectively.

Lemma 2.2 *Let Y_V be a random vector of binary variables and let A and B be two nonempty, disjoint, subsets of V. Then, for any pair $f, g \in \{c, m\}$ it holds that $Y_A \perp\!\!\!\perp Y_B$ if and only if*

$$\mathrm{cpr}_{fg}(A', A : B', B) = 1$$

for every $A' \subseteq A$ and $B' \subseteq B$ such that $A', B' \neq \varnothing$.

Proof. $Y_A \perp\!\!\!\perp Y_B$ implies $p_{fg}(A', A : B', B) = p_f(A', A)p_g(B', B)$ for every $A' \subseteq A$ and $B' \subseteq B$ and therefore, in this case, $\mathrm{cpr}_{fg}(A', A : B', B) = 1$ for every $A' \subseteq A$ and $B' \subseteq B$, as required.

We now show the reverse implication. In the case $f = c$ then $\mathrm{cpr}_{cg}(A', A : B', B) = 1$ implies that for every $\varnothing \neq A' \subseteq A$ and $\varnothing \neq B' \subseteq B$

$$\frac{p_{cg}(A', A : B', B)}{p_{cg}(\varnothing, A : B', B)} = \frac{p_{cg}(A', A : \varnothing, B)}{p_{cg}(\varnothing, A : \varnothing, B)},$$

and therefore

$$\frac{p_{cg}(A', A|B', B)}{p_{cg}(\varnothing, A|B', B)} = \frac{p_{cg}(A', A|\varnothing, B)}{p_{cg}(\varnothing, A|\varnothing, B)}. \tag{2.22}$$

The equality (2.22) trivially holds true for $A' = \varnothing$ and therefore we can sum both sides over $A' \subseteq A$ to obtain $p_{cg}(\varnothing, A|B', B)^{-1} = p_{cg}(\varnothing, A|\varnothing, B)^{-1}$, and therefore that $p_{cg}(\varnothing, A|B', B) = p_{cg}(\varnothing, A|\varnothing, B)$. Hence, the two denominators of (2.22) cancel out thereby giving $p_{cg}(A', A|B', B) = p_{cg}(A', A|\varnothing, B)$ for every $\varnothing \neq A' \subseteq A$ and $\varnothing \neq B' \subseteq B$ that implies $Y_A \perp\!\!\!\perp Y_B$ by Lemma 2.1.

Consider now the case $f = m$. The equality $\mathrm{cpr}_{mg}(A', A : B', B) = 1$ implies

$$1 = \frac{p_g(\varnothing, B)p_{mg}(A', A : B', B)}{p_g(B', B)p_{mg}(A', A : \varnothing, B)} = \frac{p_{mg}(A', A|B', B)}{p_{mg}(A', A|\varnothing, B)},$$

that, in turn, implies $p_{mg}(A', A|B', B) = p_{mg}(A', A|\varnothing, B)$. The latter, when satisfied for every $\varnothing \neq A' \subseteq A$ and $\varnothing \neq B' \subseteq B$, implies $Y_A \perp\!\!\!\perp Y_B$ by Lemma 2.1. This states the result. $\qquad\square$

The results of Lemmas 2.1 and 2.2 are feasible of a more general formulation involving conditional independence. If $C \subseteq V$ is such that $C \cap A = C \cap B = \emptyset$ then, for every $C' \subseteq C$, we can consider the conditional distribution of $Y_{A \cup B} | \{ Y_{C'} = 1, Y_{C \setminus C'} = 0 \}$ and, accordingly, define the quantities $p_{fg}(A', A : B', B | C', C)$ and $\mathrm{cpr}_{fg}(A', A : B', B | C', C)$. The formulation of Lemma 2.1 in terms of conditional independence is as follows.

Corollary 2.3 *Let Y_V be a random vector of binary variables and let A, B and C be three pairwise disjoint subsets of V such that $A, B \neq \emptyset$. Then, for any pair $f, g \in \{c, m\}$ it holds that $Y_A \perp\!\!\!\perp Y_B | Y_C$ if and only if*

$$p_{fg}(A', A | B', B : C', C) = p_{fg}(A', A | \emptyset, B : C', C)$$

for every $A' \subseteq A$ and $B' \subseteq B$ such that $A', B' \neq \emptyset$ and for every $C' \subseteq C$.

Proof. This follows from the repeated application of Lemma 2.1 for every $C' \subseteq C$. $\qquad \square$

We now give the formulation of Lemma 2.2 in terms of conditional independence.

Corollary 2.4 *Let Y_V be a random vector of binary variables and let A, B and C be three pairwise disjoint subsets of V such that $A, B \neq \emptyset$. Then, for any pair $f, g \in \{c, m\}$ it holds that $Y_A \perp\!\!\!\perp Y_B | Y_C$ if and only if*

$$\mathrm{cpr}_{fg}(A', A : B', B | C', C) = 1$$

for every $A' \subseteq A$ and $B' \subseteq B$ such that $A', B' \neq \emptyset$ and for every $C' \subseteq C$.

Proof. This follows from the repeated application of Lemma 2.2 for every $C' \subseteq C$. $\qquad \square$

The Non-binary Case

In this section the results of the binary case are extended to the general setting of non-binary variables. We show how the quantities introduced in the binary case can be defined with respect to the binary subtables identified by the elements $j \in \mathcal{J}$ and, furthermore, that the results of the previous section hold in the general polytomous case when they hold for every such binary subtable.

For every $j \in \mathcal{J}$ we define the non-binary versions of (2.9) and (2.10) as

$$p_{fg}(j_{A'}, A : j_{B'}, B) = p(Y_{A'} = j_{A'}, Y_{f(A\backslash A')} = 0, Y_{B'} = j_{B'}, Y_{g(B\backslash B')} = 0)$$

and

$$p_{fg}(j_{A'}, A | j_{B'}, B) = p(Y_{A'} = j_{A'}, Y_{f(A\backslash A')} = 0 | Y_{B'} = j_{B'}, Y_{g(B\backslash B')} = 0),$$

respectively, so that the version of Lemma 2.1 for non-binary variables is as follows.

Lemma 2.5 *Let Y_V be a random vector of categorical variables and let A and B be two nonempty, disjoint, subsets of V. Then, for any pair $f, g \in \{c, m\}$ it holds that $Y_A \perp\!\!\!\perp Y_B$ if and only if*

$$p_{fg}(j_{A'}, A | j_{B'}, B) = p_{fg}(j_{A'}, A | \varnothing, B)$$

for every $A' \subseteq A$ and $B' \subseteq B$ such that $A', B' \neq \varnothing$ and for every $j \in \mathcal{J}$.

Proof. This proof can be obtained from the proof of Lemma 2.1 by operating the following changes. (i) Replace, everywhere, $Y_{A'} = 1$ and $Y_{A''} = 1$ by $Y_{A'} = j_{A'}$ and $Y_{A''} = j_{A''}$, respectively, and, similarly, $Y_{B'} = 1$ by $Y_{B'} = j_{B'}$. (ii) Replace, everywhere, the condition *for every $\varnothing \neq A' \subseteq A$ and $\varnothing \neq B' \subseteq B$* by *for every $\varnothing \neq A' \subseteq A$, $\varnothing \neq B' \subseteq B$ and $j \in \mathcal{J}$*. (iii) Replace, twice, the sum over $A' \subseteq A$ by the sum over $j_{A'}$ for $A' \subseteq A$ and $j_{A'} \in \mathcal{J}_{A'}$, and once the sum over $A'' \subseteq A \backslash A'$ by the sum over $j_{A''}$ for $A'' \subseteq A \backslash A'$ and $j_{A''} \in \mathcal{J}_{A''}$. (iv) In the part of the proof where it is shown that the case $(f, g) = (m, m)$ implies the case $(f, g) = (c, m)$, replace $p(Y_{A'} = 1, Y_{\overline{A}\backslash\{a\}} = 0, Y_a = 1 | Y_{B'} = 1)$ by $\sum_{j_a=1}^{d_a} p(Y_{A'} = j_{A'}, Y_{\overline{A}\backslash\{a\}} = 0, Y_a = j_a | Y_{B'} = j_{B'})$ and, similarly, $p(Y_{A'} = 1, Y_{\overline{A}\backslash\{a\}} = 0, Y_a = 1)$ by $\sum_{j_a=1}^{d_a} p(Y_{A'} = j_{A'}, Y_{\overline{A}\backslash\{a\}} = 0, Y_a = j_a)$. $\qquad\square$

For every $j \in \mathcal{J}$ the non-binary versions of the cross-product ratio in (2.21) is

$$\text{cpr}_{fg}(j_{A'}, A : j_{B'}, B) = \frac{p_{fg}(\varnothing, A : \varnothing, B) \, p_{fg}(j_{A'}, A : j_{B'}, B)}{p_{fg}(\varnothing, A : j_{B'}, B) \, p_{fg}(j_{A'}, A : \varnothing, B)} \qquad (2.23)$$

so that we can give the version of Lemma 2.2 for non-binary variables.

Lemma 2.6   *Let Y_V be a random vector of categorical variables and let A and B be two nonempty, disjoint, subsets of V. Then, for any pair $f, g \in \{c, m\}$ it holds that $Y_A \perp\!\!\!\perp Y_B$ if and only if*

$$\text{cpr}_{fg}(j_{A'}, A : j_{B'}, B) = 1$$

for every $A' \subseteq A$ and $B' \subseteq B$ such that $A', B' \neq \emptyset$ and for every $j \in \mathcal{J}$.

Proof.   This proof can be obtained from the proof of Lemma 2.2 by operating the following changes. (i) Replace, everywhere, A' by $j_{A'}$ and, similarly, B' by $j_{B'}$. (ii) Replace, the condition *for every $A' \subseteq A$ and $B' \subseteq B$* by *for every $A' \subseteq A$, $B' \subseteq B$ and $j \in \mathcal{J}$.* (iii) Replace, twice, the condition *for every $\emptyset \neq A' \subseteq A$ and $\emptyset \neq B' \subseteq B$* by *for every $\emptyset \neq A' \subseteq A$, $\emptyset \neq B' \subseteq B$ and $j \in \mathcal{J}$.* (iv) Replace the sum over $A' \subseteq A$ by the sum over $j_{A'}$ for $A' \subseteq A$ and $j \in \mathcal{J}$.   □

Next, we can consider the conditional distribution of $Y_{A \cup B} | \{Y_{C'} = j_{C'}, Y_{C \setminus C'} = 0\}$ and, accordingly define the quantities

$$p_{fg}(j_{A'}, A : j_{B'}, B | j_{C'}, C) \text{ and } \text{cpr}_{fg}(j_{A'}, A : j_{B'}, B | j_{C'}, C)$$

to obtain a non-binary extension of Corollary 2.3.

Corollary 2.7   *Let Y_V be a random vector of categorical variables and let A, B and C be three pairwise disjoint subsets of V such that $A, B \neq \emptyset$. Then, for any pair $f, g \in \{c, m\}$ it holds that $Y_A \perp\!\!\!\perp Y_B | Y_C$ if and only if*

$$p_{fg}(j_{A'}, A | j_{B'}, B : j_{C'}, C) = p_{fg}(j_{A'}, A | \emptyset, B : j_{C'}, C)$$

for every $A' \subseteq A, B' \subseteq B$ and $C' \subseteq C$, such that $A', B' \neq \emptyset$, and for every $j \in \mathcal{J}$.

Proof.   This follows from the repeated application of Lemma 2.5 for every $C' \subseteq C$ and $j \in \mathcal{J}$.   □

Finally, we give the non-binary version of Corollary 2.4.

Corollary 2.8   *Let Y_V be a random vector of categorical variables and let A, B and C be three pairwise disjoint subsets of V such that $A, B \neq \emptyset$. Then, for any pair $f, g \in \{c, m\}$ it holds that $Y_A \perp\!\!\!\perp Y_B | Y_C$ if and only if*

$$\text{cpr}_{fg}(j_{A'}, A : j_{B'}, B | j_{C'}, C) = 1$$

for every $A' \subseteq A$, $B' \subseteq B$ and $C' \subseteq C$, such that $A', B' \neq \emptyset$, and for every $j \in \mathcal{J}$.

Proof. This follows from the repeated application of Lemma 2.6 for every $C' \subseteq C$ and $j \in \mathcal{J}$. $\qquad\qquad\qquad\qquad\qquad\qquad\qquad\quad\square$

3
Möbius Inversion

The general statement of Möbius inversion formula was first given by Weisner (1935) and Hall (1934) in the area of group theory. Subsequently, in a fundamental paper on Möbius functions, Rota (1964) showed the importance of this theory in combinatorial mathematics. Since then, the theory of Möbius inversion and related topics has become an active area of combinatorics. We devote one full chapter to this topic because the material given here makes it possible to present in a common framework some fundamental features of all the families of graphical models considered in the forthcoming chapters. Furthermore, Möbius inversion turns out to be central also in the exponential family representation of undirected graph models as well as for the derivation of the asymptotic theory for these models.

3.1 Preliminaries

3.1.1 Notation and Terminology

Here we extend the notation used so far to include vectors and matrices with entries indexed by the subsets of V. More specifically, we write $\mathcal{P}(V)$ to denote the power set of V, $\mathcal{P}(V) = \{D : D \subseteq V\}$, and denote by $\theta_{\mathcal{P}(V)} = (\theta_D)_{D \subseteq V}$ a real vector indexed by the subsets of V. Hence, any subset $\mathcal{A} \subseteq \mathcal{P}(V)$ identifies a subvector $\theta_{\mathcal{A}} = (\theta_D)_{D \in \mathcal{A}}$ of $\theta_{\mathcal{P}(V)}$. We omit the subscript for vectors indexed by the whole power set of V, thereby setting $\theta = \theta_{\mathcal{P}(V)}$. Likewise, the product set $\mathcal{P}(V) \times \mathcal{P}(V)$ indexes the entries of the matrix $\mathbb{G}_{\mathcal{P}(V)\mathcal{P}(V)} = (\mathbb{G}_{D,E})_{D,E \subseteq V}$ where the first argument indexes the rows and the second the columns. Any subset $A \subseteq V$ identifies a submatrix $\mathbb{G}_{\mathcal{P}(A)\mathcal{P}(A)}$ of $\mathbb{G}_{\mathcal{P}(V)\mathcal{P}(V)}$ and, more generally, we denote by $\mathbb{G}_{\mathcal{A}\mathcal{B}}$ the submatrix with rows and columns indexed by the elements of the subsets $\mathcal{A}, \mathcal{B} \subseteq \mathcal{P}(V)$, respectively. In order to lighten notation we set $\mathbb{G}_{\mathcal{A}} = \mathbb{G}_{\mathcal{A}\mathcal{A}}$, $\mathbb{G}_A = \mathbb{G}_{\mathcal{P}(A)}$ and $\mathbb{G} = \mathbb{G}_V$.

We will mostly deal with subsets of $\mathcal{P}(V)$ that are simplicial complexes. Formally, a subset $\mathcal{C} \subseteq \mathcal{P}(V)$ is called an *(abstract) simplicial complex* if

$E \in C$ and $D \subseteq E$ implies $D \in C$. We remark that $\emptyset \in C$ and write $C_{\&}$ to denote that C has been deprived of the empty set; formally $C_{\&} = C \setminus \emptyset$.

The elements of the power set of V can be partially ordered by assuming that, for $A, B \subseteq V$, the set A precedes the set B whenever $A \subseteq B$. Hereafter, when we say that a matrix is either diagonal or triangular, we mean that this property is fulfilled under any total ordering of its rows and columns compatible with the above partial ordering.

3.1.2 The Zeta and the Möbius Matrices

The *zeta matrix*, denoted by $\mathbb{Z}$, and its inverse, called the *Möbius matrix*, denoted by $\mathbb{M}$, are almost ubiquitous in this text. They are repeatedly used in the presentation of several topics as well as in many of the proofs. It is thus worthwhile to present them in a detailed way. The relevance of these two matrices rests in the fact that they result from the implementation of the Möbius inversion formula (see e.g., Speed, 1983; Lovász, 1993; Jokinen, 2006).

We start from the $\mathbb{Z}$ and $\mathbb{M}$ matrices associated with singleton sets. Hence, for every entry $v \in V$ the 2×2 matrices $\mathbb{Z}_v$ and $\mathbb{M}_v$ indexed by $\mathcal{P}(\{v\})$ are defined as

$$\begin{array}{cc} & \emptyset \quad \{v\} \\ \begin{array}{c} \emptyset \\ \{v\} \end{array} & \begin{pmatrix} 1 & 1 \\ 0 & 1 \end{pmatrix} = \mathbb{Z}_v \end{array} \quad \text{and} \quad \begin{array}{cc} & \emptyset \quad \{v\} \\ \begin{array}{c} \emptyset \\ \{v\} \end{array} & \begin{pmatrix} 1 & -1 \\ 0 & 1 \end{pmatrix} = \mathbb{M}_v \end{array}$$

where we have also displayed the indexes associated with every row and column. We now consider an arbitrary, nonempty, subset $A \subseteq V$. The zeta and the Möbius matrices *associated with* A are indexed by the elements of $\mathcal{P}(A)$ and given by

$$\mathbb{Z}_A = \otimes_{v \in A} \mathbb{Z}_v \quad \text{and} \quad \mathbb{M}_A = \otimes_{v \in A} \mathbb{M}_v,$$

respectively, where $\otimes$ denotes the Kronecker product of matrices. It is easy to see that $\mathbb{Z}_v^{-1} = \mathbb{M}_v$ and, consequently, an immediate consequence of the properties of the Kronecker product is that

$$\mathbb{Z}_A^{-1} = \otimes_{v \in A} \mathbb{Z}_v^{-1} = \otimes_{v \in A} \mathbb{M}_v = \mathbb{M}_A. \tag{3.1}$$

Every pair of subsets $D, E \subseteq A$ uniquely identifies the entry

$$(\mathbb{Z}_A)_{D,E} = \prod_{v \in A} (\mathbb{Z}_v)_{D \cap \{v\}, E \cap \{v\}} \tag{3.2}$$

of $\mathbb{Z}_A$, and similarly for $\mathbb{M}_A$. On the other hand, every entry (3.2) can be obtained as a direct function of the indexing subsets.

Proposition 3.1 *Let $\mathbb{Z}_A$ and $\mathbb{M}_A$ be the zeta and Möbius matrices, respectively, associated with the nonempty subset $A \subseteq V$. Then, for every $D, E \subseteq A$*

$$(\mathbb{Z}_A)_{D,E} = 1(D \subseteq E) \quad and \quad (\mathbb{M}_A)_{D,E} = (-1)^{|E \setminus D|} 1(D \subseteq E),$$

where $1(\cdot)$ denotes the indicator function.

Proof. It follows from (3.2) that the entry $(\mathbb{Z}_A)_{D,E}$ of $\mathbb{Z}_A$ is equal either to zero or to one and, more specifically, it is equal to one if and only if

$$(\mathbb{Z}_v)_{D \cap \{v\}, E \cap \{v\}} = 1 \text{ for every } v \in A$$

that is, if and only if there exists no $v \in A$ such that both $v \in D$ and $v \notin E$. Clearly, this happens if and only if $D \subseteq E$.

We now turn to $\mathbb{M}_A$. It follows from (3.2) that (i) every entry $(\mathbb{M}_A)_{D,E}$ is equal either to zero or to plus/minus one and, (ii) because $\mathbb{Z}_v$ and $\mathbb{M}_v$ have the same entries equal to zero, then also $(\mathbb{M}_A)_{D,E} = 0$ if and only if $(\mathbb{Z}_A)_{D,E} = 0$, so that we can write $(\mathbb{M}_A)_{D,E} = \pm 1(D \subseteq E)$. Finally, for every $v \in A$, it holds that $(\mathbb{M}_v)_{D \cap \{v\}, E \cap \{v\}} = -1$ if and only if $D \cap \{v\} = \varnothing$ and $E \cap \{v\} = \{v\}$; i.e., if and only if $v \in E \setminus D$. It follows that if $(\mathbb{M}_A)_{D,E} = \pm 1$ then it is equal to -1 if and only if $|E \setminus D|$ is odd so that we can write $(\mathbb{M}_A)_{D,E} = (-1)^{|E \setminus D|} 1(D \subseteq E)$ as required. $\square$

For example, the matrix $\mathbb{Z}_{\{a,b,c\}}$, displayed with its row and column indexes, can be written as

	$\varnothing$	$\{a\}$	$\{b\}$	$\{c\}$	$\{a,b\}$	$\{a,c\}$	$\{b,c\}$	$\{a,b,c\}$
$\varnothing$	1	1	1	1	1	1	1	1
$\{a\}$	0	1	0	0	1	1	0	1
$\{b\}$	0	0	1	0	1	0	1	1
$\{c\}$	0	0	0	1	0	1	1	1
$\{a,b\}$	0	0	0	0	1	0	0	1
$\{a,c\}$	0	0	0	0	0	1	0	1
$\{b,c\}$	0	0	0	0	0	0	1	1
$\{a,b,c\}$	0	0	0	0	0	0	0	1

whereas $\mathbb{M}_{\{a,b,c\}}$ is

$$
\begin{array}{c}
 \\
\varnothing \\
\{a\} \\
\{b\} \\
\{c\} \\
\{a,b\} \\
\{a,c\} \\
\{b,c\} \\
\{a,b,c\}
\end{array}
\begin{array}{cccccccc}
\varnothing & \{a\} & \{b\} & \{c\} & \{a,b\} & \{a,c\} & \{b,c\} & \{a,b,c\} \\
\left(\begin{array}{cccccccc}
1 & -1 & -1 & -1 & 1 & 1 & 1 & -1 \\
0 & 1 & 0 & 0 & -1 & -1 & 0 & 1 \\
0 & 0 & 1 & 0 & -1 & 0 & -1 & 1 \\
0 & 0 & 0 & 1 & 0 & -1 & -1 & 1 \\
0 & 0 & 0 & 0 & 1 & 0 & 0 & -1 \\
0 & 0 & 0 & 0 & 0 & 1 & 0 & -1 \\
0 & 0 & 0 & 0 & 0 & 0 & 1 & -1 \\
0 & 0 & 0 & 0 & 0 & 0 & 0 & 1
\end{array}\right)
\end{array}.
$$

Notice that if one respects the order relation "$\subseteq$" in the rows and columns of the matrices, then both $\mathbb{Z}_A$ and $\mathbb{M}_A$ are upper triangular with diagonal entries equal to one. Furthermore, it follows from Proposition 3.1 that it makes sense to set $\mathbb{Z}_\varnothing = \mathbb{M}_\varnothing = 1$, and that if $B \subseteq A$ then the zeta and Möbius matrices associated with B, i.e., $\mathbb{Z}_B$ and $\mathbb{M}_B$, are equal to the relevant submatrices of $\mathbb{Z}_A$ and $\mathbb{M}_A$, respectively. Finally, we recall that we may omit the subscript when it is equal to V, so that $\mathbb{Z} = \mathbb{Z}_V$ and $\mathbb{M} = \mathbb{M}_V$.

3.2 The Möbius Inversion Formula

Here we provide the Möbius inversion formula in a very simplified setting, whereas Section 3.3 deals with a slightly more general version of this result. Applications of Möbius inversion to the analysis of discrete graphical models can be found, among others, in Lauritzen (1996); Drton and Richardson (2008a); Drton (2009); Massam et al. (2009); La Rocca and Roverato (2017); Lupparelli and Roverato (2017).

Theorem 3.2 *Let $\theta = (\theta_D)_{D \subseteq V}$ and $\omega = (\omega_D)_{D \subseteq V}$ be two real vectors indexed by the subsets of a finite set V, then it holds that*

$$
\omega_D = \sum_{D' \subseteq D} \theta_{D'} \quad \text{if and only if} \quad \theta_D = \sum_{D' \subseteq D} (-1)^{|D \setminus D'|} \omega_{D'}, \quad (3.3)
$$

where identities are intended for all $D \subseteq V$. This statement can be written in matrix notation as

$$\theta = \mathbb{M}^T \omega \quad \text{if and only if} \quad \omega = \mathbb{Z}^T \theta.$$

The proof of this theorem is straightforward if one approaches it from the matrix formulation because $\mathbb{Z} = \mathbb{M}^{-1}$, as shown in (3.1). The two equivalent formulations of the Möbius inversion formula given in Theorem 3.2 are not redundant. Indeed, they provide a powerful tool that will be used throughout this text to derive a compact matrix formulation of the quantities of interest as well as an explicit description of the individual entries which constitute such quantities.

3.2.1 Two Basic Lemmas

The following lemmas provide two properties of Möbius inversion, which will be useful in connection with the cross-product ratios of Corollary 2.4 (see also La Rocca and Roverato, 2017).

Lemma 3.3 *Let $\theta = (\theta_D)_{D \subseteq V}$ and $\omega = (\omega_D)_{D \subseteq V}$ be two real vectors indexed by the subsets of a finite set V. For a term A, B, C of mutually disjoint subsets of V the following statements are equivalent*

(i) *for every $D \subseteq A \cup B \cup C$ such that both $D \cap A \neq \emptyset$ and $D \cap B \neq \emptyset$ it holds that*

$$\theta_D = 0,$$

(ii) *for every $A' \subseteq A$ and $B' \subseteq B$ such that $A', B' \neq \emptyset$, and for every $C' \subseteq C$ it holds that*

$$\omega_{A' \cup B' \cup C'} - \omega_{A' \cup C'} - \omega_{B' \cup C'} + \omega_{C'} = 0.$$

Proof. (i) $\Rightarrow$ (ii). Because for every $D \subseteq V$ it holds that $\omega_D = \sum_{D' \subseteq D} \theta_{D'}$, then for every $\emptyset \neq A' \subseteq A$, $\emptyset \neq B' \subseteq B$ and $C' \subseteq C$ we can write

$$\omega_{A' \cup B' \cup C'} = \sum_{D' \subseteq A' \cup B' \cup C'} \theta_{D'}, \tag{3.4}$$

and it follows from (i) that $\theta_{D'} = 0$ for every D' such that both $D' \not\subseteq A' \cup C'$ and $D' \not\subseteq B' \cup C'$. Hence, we can write (3.4) as

$$\omega_{A' \cup B' \cup C'} = \sum_{D' \subseteq A' \cup C'} \theta_{D'} + \sum_{D' \subseteq B' \cup C'} \theta_{D'} - \sum_{D' \subseteq C'} \theta_{D'}$$

$$= \omega_{A' \cup C'} + \omega_{B' \cup C'} - \omega_{C'} \tag{3.5}$$

which implies (ii). Note that the term $-\sum_{D' \subseteq C'} \theta_{D'}$ in (3.5) is necessary because, otherwise, every term $\theta_{D'}$ with $D' \in C'$ would be considered twice, once for each of the other sums in (3.5).

(ii) $\Rightarrow$ (i). We prove this result by induction on the cardinality of the set $A \cup B \cup C$.

We first show that the result is true for $|A \cup B \cup C| = 2$. In this case, $A = \{a\}$, $B = \{b\}$ and $C = \emptyset$ because A and B cannot be empty. Hence, we have to show that $\omega_{\{a,b\}} - \omega_a - \omega_b + \omega_\emptyset = 0$ implies $\theta_D = \theta_{\{a,b\}} = 0$ and the result follows immediately from the fact that

$$\theta_{\{a,b\}} = \sum_{D' \subseteq \{a,b\}} (-1)^{|\{a,b\} \setminus D'|} \omega_{D'}$$

$$= \omega_{\{a,b\}} - \omega_a - \omega_b + \omega_\emptyset.$$

We now set $k > 2$ and show that if the result holds for $|A \cup B \cup C| < k$ then it also holds true for $|A \cup B \cup C| = k$. More specifically, the induction assumption implies that $\theta_{D'} = 0$ for every $D' \subset A \cup B \cup C$ such that $D' \cap A \neq \emptyset$ and $D' \cap B \neq \emptyset$. This implies both that it is sufficient to show that (ii) implies $\theta_D = \theta_{A \cup B \cup C} = 0$ and that

$$\sum_{D' \subseteq A \cup B \cup C} \theta_{D'} = \theta_{A \cup B \cup C} + \sum_{D' \subseteq A \cup C} \theta_{D'} + \sum_{D' \subseteq B \cup C} \theta_{D'} - \sum_{D' \subseteq C} \theta_{D'}. \tag{3.6}$$

Every sum in (3.6) corresponds to an entry of ω as follows $\omega_{A \cup B \cup C} = \theta_{A \cup B \cup C} + \omega_{A \cup C} + \omega_{B \cup C} - \omega_C$, so that $\theta_{A \cup B \cup C} = \omega_{A \cup B \cup C} - \omega_{A \cup C} - \omega_{B \cup C} + \omega_C$ and because, by (ii), it holds that $\omega_{A \cup B \cup C} - \omega_{A \cup C} - \omega_{B \cup C} + \omega_C = 0$, it follows that $\theta_{A \cup B \cup C} = 0$ as required. $\qquad \square$

For a subset $B \subseteq V$ we set $A = V \setminus A$ and denote by $s_B(\omega) = (\omega_{B \cup A'})_{A' \subseteq A}$ the subvector of ω whose entries are indexed by supersets of B. In this context, an object of interest is the vector $\theta_{[B]} = \mathbb{M}_A^T s_B(\omega)$, with entries

$$\theta_{[B]A'} = \sum_{A'' \subseteq A'} (-1)^{|A' \setminus A''|} \omega_{B \cup A''} \quad \text{for every } A' \subseteq A. \tag{3.7}$$

If $\theta = \mathbb{M}^T \omega$ then it follows immediately from the definition of $\theta_{[B]}$ that $\theta_{[\varnothing]} = \theta$ but, in general, $\theta_{[B]}$ is not a subvector of θ, and the following lemma states the connection existing between $\theta_{[B]}$ and θ (see also La Rocca and Roverato, 2017).

Lemma 3.4 *For a real vector ω indexed by the power set of V let $\theta = \mathbb{M}^T \omega$. Furthermore, for a subset $B \subseteq V$ let $\theta_{[B]}$ be the vector defined by (3.7), where $A = V \setminus B$. Then it holds that*

$$\theta_{[B]A'} = \sum_{B' \subseteq B} \theta_{A' \cup B'} \quad \text{for every } A' \subseteq A. \tag{3.8}$$

Proof. If we let $\theta^* = (\theta^*_{A'})_{A' \subseteq A}$ be the vector with entries $\theta^*_{A'} = \sum_{B' \subseteq B} \theta_{A' \cup B'}$ then, in order to establish (3.8), we have to show that $\theta_{[B]} = \theta^*$. Because $\omega = \mathbb{Z}^T \theta$, then for every $A' \subseteq A$

$$\omega_{B \cup A'} = \sum_{D \subseteq B \cup A'} \theta_D = \sum_{A'' \subseteq A'} \sum_{B' \subseteq B} \theta_{A'' \cup B'} = \sum_{A'' \subseteq A'} \theta^*_{A''}, \tag{3.9}$$

and (3.9) can be written in vector form as $s_B(\omega) = \mathbb{Z}_A^T \theta^*$. On the other hand, by definition $\theta_{[B]} = \mathbb{M}_A^T s_B(\omega)$, so it also holds that $s_B(\omega) = \mathbb{Z}_A^T \theta_{[B]}$. Hence, it holds that $\mathbb{Z}_A^T \theta_{[B]} = \mathbb{Z}_A^T \theta^*$ and, as $\mathbb{Z}_A$ has full rank, we have $\theta_{[B]} = \theta^*$, as required. $\square$

3.3 Möbius Inversion and Partially Ordered Sets

This section deals with more advanced topics on Möbius inversion, which are mostly used in the asymptotic theory presented in Section 4.11. Readers may prefer to skip this section on first reading and refer back to it when the relevant results are required.

Let $(\mathcal{P}, \leq)$ be a finite *partially ordered set*, that is a finite set $\mathcal{P}$ endowed with the partial ordering "$\leq$." Furthermore, let $f(\cdot)$ be a real function on $\mathcal{P}$. We consider an arbitrary total ordering of the elements of $\mathcal{P}$, compatible with the partial ordering "$\leq$," so that $f(\cdot)$ is univocally associated with the vector, indexed by $\mathcal{P}$, $f_{\mathcal{P}} = \{f(D)\}_{D \in \mathcal{P}}$. Similarly, any real function $g(\cdot, \cdot)$ on $\mathcal{P} \times \mathcal{P}$ is univocally associated with a matrix $\mathbb{G}_{\mathcal{P}} = \{g(D, H)\}_{D, H \in \mathcal{P}}$.

For any finite partially ordered set $(\mathcal{P}, \leq)$ there are two important numerical functions defined on $\mathcal{P} \times \mathcal{P}$: the *zeta function* $z_{\mathcal{P}}(\cdot, \cdot)$ given by $z_{\mathcal{P}}(D, H) = 1$ if $D \leq H$, and 0 otherwise, and the *Möbius function* $m_{\mathcal{P}}(\cdot, \cdot)$ defined as

$$m_{\mathcal{P}}(D, H) = \begin{cases} 1 & \text{if } D = H; \\ -\sum_{D \leq F < H} m_{\mathcal{P}}(D, F) & \text{if } D < H; \\ 0 & \text{otherwise.} \end{cases}$$

We refer to Rota (1964) for a more detailed account on the theory of Möbius functions. The matrix associated with the zeta function is called the *zeta matrix* and denoted by $\mathbb{Z}_{\mathcal{P}}$. It is straightforward to see that $\mathbb{Z}_{\mathcal{P}}$ is upper triangular with all diagonal entries equal to one, so that $\mathbb{Z}_{\mathcal{P}}$ is invertible with determinant equal to one; $|\mathbb{Z}_{\mathcal{P}}| = 1$. The matrix associated with the Möbius function, denoted by $\mathbb{M}_{\mathcal{P}}$, is called the *Möbius matrix*. It should be clear that the zeta and Möbius matrices defined here are generalizations of the matrices of Section 3.1.2. It can be checked that, also in this more general setting, the product $\mathbb{M}_{\mathcal{P}} \times \mathbb{Z}_{\mathcal{P}}$ is equal to the identity matrix, which implies that $\mathbb{M}_{\mathcal{P}} = \mathbb{Z}_{\mathcal{P}}^{-1}$ and, consequently, that $\mathbb{M}_{\mathcal{P}}$ is upper triangular with diagonal entries equal to one; see Speed (1983), Lovász (1993, p. 216) and Lütkepol (1996, p. 164). We can now give a more general version of the Möbius inversion formula that deals with arbitrary partially ordered sets.

Theorem 3.5 *Consider a finite partially ordered set $\mathcal{P}$ and let $f(\cdot)$ and $h(\cdot)$ be two real functions on $\mathcal{P}$. Then*

(a) $h(D) = \displaystyle\sum_{E \geq D} f(E)$

   if and only if

   $f(D) = \displaystyle\sum_{E \in \mathcal{P}} m_{\mathcal{P}}(D, E) h(E),$

(b) $h(D) = \displaystyle\sum_{E \leq D} f(E)$

   if and only if

   $f(D) = \displaystyle\sum_{E \in \mathcal{P}} m_{\mathcal{P}}(E, D) h(E),$

where identities are intended for all $D \in \mathcal{P}$. These two statements can be rewritten in matrix notation as follows,

(A) $h_{\mathcal{P}} = \mathbb{Z}_{\mathcal{P}}\,f_{\mathcal{P}}$ *if and only if* $f_{\mathcal{P}} = \mathbb{M}_{\mathcal{P}}\,h_{\mathcal{P}},$

(B) $h_{\mathcal{P}} = \mathbb{Z}_{\mathcal{P}}^{T}\,f_{\mathcal{P}}$ *if and only if* $f_{\mathcal{P}} = \mathbb{M}_{\mathcal{P}}^{T}\,h_{\mathcal{P}}.$

As well as for Theorem 3.2, also the proof of Theorem 3.5 is straightforward if one approaches it from the matrix formulation in (A) and (B) because $\mathbb{M}_{\mathcal{P}} = \mathbb{Z}_{\mathcal{P}}^{-1}$; see also Lovász (1993, p. 218) and Rota (1964, proposition 2 and corollary 1).

We now consider a finite set V and turn to the partially ordered set $(\mathcal{C}, \subseteq)$ where $\mathcal{C} \subseteq \mathcal{P}(V)$ is a simplicial complex. In this case, the zeta and Möbius functions can be written as

$$z_{\mathcal{C}}(D, H) = 1(D \subseteq H) \quad \text{and} \quad m_{\mathcal{C}}(D, H) = (-1)^{|H \setminus D|} \times 1(D \subseteq H)$$
(3.10)

respectively (see Lovász, 1993, p. 216). Here, it deserves notice that if $\mathbb{Z}$ is the zeta matrix associated with $\mathcal{P}(V)$, then the zeta matrix $\mathbb{Z}_{\mathcal{C}}$ associated with a simplicial complex $\mathcal{C} \subseteq \mathcal{P}(V)$ is the submatrix of $\mathbb{Z}$ indexed by $\mathcal{C} \times \mathcal{C}$. Furthermore, the Möbius matrix associated with $\mathcal{C}$ can be obtained in two different but equivalent ways, namely as the inverse of the zeta matrix, $\mathbb{M}_{\mathcal{C}} = \mathbb{Z}_{\mathcal{C}}^{-1}$, and as the submatrix of $\mathbb{M} = \mathbb{Z}^{-1}$ indexed by $\mathcal{C} \times \mathcal{C}$. When $\mathcal{P}$ is a simplicial complex, Theorem 3.5 can be restated as follows.

Corollary 3.6 *For a finite set V consider a simplicial complex $\mathcal{C} \subseteq \mathcal{P}$ and let $f(\cdot)$ and $h(\cdot)$ be two real functions on $\mathcal{P}$. Then*

(a') $h(D) = \sum_{E \subseteq V \setminus D} f(F)\, 1(F \in \mathcal{C})$

if and only if

$f(D) = \sum_{E \subseteq V \setminus D} (-1)^{|E|}\, h(F)\, 1(F \in \mathcal{C}),$

(b') $h(D) = \sum_{E \subseteq D} f(E)$

if and only if

$f(D) = \sum_{E \subseteq D} (-1)^{|D \setminus E|} h(E);$

where $F = D \cup E$ and identities are intended for all $D \in \mathcal{C}$. The matrix version of (a') can be obtained from (A) of Theorem 3.6 by replacing $\mathcal{P}$ with $\mathcal{C}$ everywhere, and the matrix version of (b') can be obtained in the same way from (B).

Proof. We have to show that for the simplicial complex $\mathcal{C}$ the four identities in (a) and (b) of Theorem 3.5 are equivalent to the corresponding identities in (a') and (b'). Firstly, we notice that for any function $g(\cdot)$ on $\mathcal{C}$ it holds that

$$\sum_{E \geq D} g(E) = \sum_{E \subseteq V \setminus D} g(D \cup E) \, \mathbb{1}(D \cup E \in \mathcal{C}) = \sum_{E \subseteq V \setminus D} g(F) \, \mathbb{1}(F \in \mathcal{C})$$

$$(3.11)$$

where $F = D \cup E$. The equivalence of the identities on the left-hand side of (a) and (a') is an immediate consequence of (3.11):

$$h(D) = \sum_{E \geq D} f(E) = \sum_{E \subseteq V \setminus D} f(F) \, \mathbb{1}(F \in \mathcal{C}).$$

Consider now the identity on the right-hand side of (a). By (3.10)

$$f(D) = \sum_{E \in \mathcal{C}} m_{\mathcal{C}}(D, E) h(E) = \sum_{E \in \mathcal{C}} (-1)^{|E \setminus D|} \, \mathbb{1}(D \subseteq E) \, h(E)$$

and as $D \subseteq E$ if and only if $E \geq D$, then it holds that

$$f(D) = \sum_{E \geq D} (-1)^{|E \setminus D|} h(E)$$

so that we can apply (3.11) to obtain the right-hand side of (a')

$$f(D) = \sum_{E \subseteq V \setminus D} (-1)^{|(D \cup E) \setminus D|} h(D \cup E) \mathbb{1}(D \cup E \in \mathcal{C})$$

$$= \sum_{E \subseteq V \setminus D} (-1)^{|E|} h(F) \mathbb{1}(F \in \mathcal{C}).$$

The equivalence of the identities on the left-hand side of (b) and (b') is straightforward, whereas to show the equivalence of the identities on the right-hand side (b) and (b') it is sufficient to apply (3.10)

$$f(D) = \sum_{E \in \mathcal{P}} m_{\mathcal{P}}(E, D) h(E) = \sum_{E \in \mathcal{C}} (-1)^{|D \setminus E|} \mathbb{1}(E \subseteq D) h(E)$$

$$= \sum_{E \subseteq D} (-1)^{|D \setminus E|} h(E);$$

and the proof is complete. $\square$

We now turn to the bivariate case. For a simplicial complex $\mathcal{C}$, the product space $\mathcal{C} \times \mathcal{C}$ is a partially ordered set naturally endowed with the ordering

$$(D, D') \leq (H, H') \text{ if and only if } D \subseteq H \text{ and } D' \subseteq H'. \tag{3.12}$$

It follows from (3.12) and (3.10) that

$$
\begin{aligned}
z_{\mathcal{C} \times \mathcal{C}}\{(D, D'), (H, H')\} &= z_{\mathcal{C}}(D, H) \times z_{\mathcal{C}}(D', H') \\
&= 1(D \subseteq H) \times 1(D' \subseteq H') \tag{3.13}
\end{aligned}
$$

and therefore that the corresponding zeta matrix is $\mathbb{Z}_{\mathcal{C} \times \mathcal{C}} = \mathbb{Z}_{\mathcal{C}} \otimes \mathbb{Z}_{\mathcal{C}}$. The Möbius matrix is $\mathbb{M}_{\mathcal{C} \times \mathcal{C}} = \mathbb{Z}_{\mathcal{C} \times \mathcal{C}}^{-1} = \mathbb{M}_{\mathcal{C}} \otimes \mathbb{M}_{\mathcal{C}}$, so that the Möbius function of $\mathcal{C} \times \mathcal{C}$ becomes

$$
\begin{aligned}
m_{\mathcal{C} \times \mathcal{C}}\{(D, D'), (H, H')\} &= m_{\mathcal{C}}(D, H) \times m_{\mathcal{C}}(D', H') \\
&= (-1)^{|H \setminus D| + |H' \setminus D'|} \times 1(D \subseteq H) \\
&\quad \times 1(D' \subseteq H'), \tag{3.14}
\end{aligned}
$$

see Rota (1964, proposition 5). In this case, Theorem 3.5 can be applied with respect to two functions on $\mathcal{C} \times \mathcal{C}$ by setting $\mathcal{P} = \mathcal{C} \times \mathcal{C}$ and of special interest is the matrix formulation of the Möbius inversion formula.

Corollary 3.7 *For a finite set V consider a simplicial complex $\mathcal{C} \subseteq \mathcal{P}$ and let $g(\cdot, \cdot)$ and $k(\cdot, \cdot)$ be functions on $\mathcal{C} \times \mathcal{C}$ with associated matrices $\mathbb{G}_{\mathcal{C}}$ and $\mathbb{K}_{\mathcal{C}}$, respectively.*

Then

$$(a'') \quad k(D, D') = \sum_{(E, E') \geq (D, D')} g(E, E')$$

if and only if

$$g(D, D') = \sum_{(E, E') \in \mathcal{C} \times \mathcal{C}} m_{\mathcal{C} \times \mathcal{C}}\{(D, D'), (E, E')\} k(E, E')$$

$$(b'') \quad k(D, D') = \sum_{(E, E') \leq (D, D')} g(E, E')$$

if and only if

$$g(D, D') = \sum_{(E, E') \in \mathcal{C} \times \mathcal{C}} m_{\mathcal{C} \times \mathcal{C}}\left\{(E, E'), (D, D')\right\} k(E, E')$$

where identities are intended for all $(D, D') \in C \times C$ and the zeta and Möbius functions are given in (3.13) and (3.14), respectively. These two statements can be rewritten in a matrix notation as follows,

(A'') $\mathbb{K}_C = \mathbb{Z}_C \mathbb{G}_C \mathbb{Z}_C^T$ *if and only if* $\mathbb{G}_C = \mathbb{M}_C \mathbb{K}_C \mathbb{M}_C^T$;

(B'') $\mathbb{K}_C = \mathbb{Z}_C^T \mathbb{G}_C \mathbb{Z}_C$ *if and only if* $\mathbb{G}_C = \mathbb{M}_C^T \mathbb{K}_C \mathbb{M}_C$.

Proof. Points (a'') and (b'') are a straightforward application of Theorem 3.5 obtained by replacing $\mathcal{P}$ with $C \times C$, $h(\cdot)$ with $k(\cdot, \cdot)$, $f(\cdot)$ with $g(\cdot, \cdot)$, D with (D, D') and, finally, E with (E, E'). We have now to show that (A'') and (B'') are equivalent to (a'') and (b''), respectively. To improve readability we split the proof into smaller parts.

1. Let $\mathrm{vec}(\mathbb{G}_C^T)$ be the $|C|^2 \times 1$ vector obtained by stacking the (transposed) rows of $\mathbb{G}_C$ into a single column, and similarly for $\mathrm{vec}(\mathbb{K}_C^T)$. The entries of these two vectors follow a common ordering and here we show that such ordering is compatible with the partial ordering of $C \times C$. Consider, without loss of generality, $\mathrm{vec}(\mathbb{G}_C^T)$. Two pairs $(D, D'), (H, H') \in C \times C$ index the entries $\mathbb{G}_{D,D'}$ and $\mathbb{G}_{H,H'}$ of $\mathbb{G}$, respectively, and we have to show that if $(D, D') \leq (H, H')$ then $\mathbb{G}_{D,D'}$ either is equal or precedes $\mathbb{G}_{H,H'}$ in $\mathrm{vec}(\mathbb{G}_C^T)$. This is an immediate consequence of the fact that $\mathrm{vec}(\mathbb{G}_C^T)$ is the vector made up by taking the entries of $\mathbb{G}$ by row, and that $(D, D') \leq (H, H')$ implies both $D \leq H$ and $D' \leq H'$. Because the rows and columns of $\mathbb{G}_C$ follow a common ordering compatible with the partial ordering of C, then the row indexed by D either is equal or precede the row indexed by H and the column indexed by D' either is equal or precede the column indexed by H'. Thus the ordering of the entries of $\mathrm{vec}(\mathbb{G}_C^T)$ is compatible with the partial ordering of $C \times C$.

2. Consider the matrices $\mathbb{Z}_C \otimes \mathbb{Z}_C$ and $\mathbb{M}_C \otimes \mathbb{M}_C$. Because the rows and columns of $\mathbb{Z}_C$ and $\mathbb{M}_C$ follow the same ordering as the rows and columns of $\mathbb{G}_C$ and $\mathbb{K}_C$, it is an immediate consequence of the definition of the Kronecker product that the rows and columns of $\mathbb{Z}_C \otimes \mathbb{Z}_C$ and $\mathbb{M}_C \otimes \mathbb{M}_C$ follow the same ordering as the entries of $\mathrm{vec}(\mathbb{G}_C^T)$ and $\mathrm{vec}(\mathbb{K}_C^T)$.

3. Points (a'') and (b'') are equivalent to

  (i) $\mathrm{vec}(\mathbb{K}_C^T) = (\mathbb{Z}_C \otimes \mathbb{Z}_C)\, \mathrm{vec}(\mathbb{G}_C^T)$

  if and only if

  $\mathrm{vec}(\mathbb{G}_C^T) = (\mathbb{M}_C \otimes \mathbb{M}_C)\, \mathrm{vec}(\mathbb{K}_C^T)$.

(ii) $\mathrm{vec}(\mathbb{K}_{\mathcal{C}}^T) = (\mathbb{Z}_{\mathcal{C}}^T \otimes \mathbb{Z}_{\mathcal{C}}^T)\, \mathrm{vec}(\mathbb{G}_{\mathcal{C}}^T)$

if and only if

$\mathrm{vec}(\mathbb{G}_{\mathcal{C}}^T) = (\mathbb{M}_{\mathcal{C}}^T \otimes \mathbb{M}_{\mathcal{C}}^T)\, \mathrm{vec}(\mathbb{K}_{\mathcal{C}}^T).$

respectively. This can be obtained by direct application of Theorem 3.5 by setting $\mathcal{P} = \mathcal{C} \times \mathcal{C}$, $\mathbb{Z}_{\mathcal{P}} = \mathbb{Z}_{\mathcal{C}} \otimes \mathbb{Z}_{\mathcal{C}}$, $\mathbb{M}_{\mathcal{P}} = \mathbb{M}_{\mathcal{C}} \otimes \mathbb{M}_{\mathcal{C}}$, $h_{\mathcal{P}} = \mathrm{vec}(\mathbb{K}_{\mathcal{C}}^T)$ and $f_{\mathcal{P}} = \mathrm{vec}(\mathbb{G}_{\mathcal{C}}^T)$. Points 1 and 2 above guarantee that the entries of all the matrices and vectors involved in (i) and (ii) follow a common ordering compatible with the partial ordering of $\mathcal{C} \times \mathcal{C}$.

4. Finally, we show that (A'') and (B'') are equivalent to (i) and (ii) of the previous point, respectively. Consider the first identity in (i). By applying the rule for the Kronecker product described, for instance, in Lütkepol (1996, p. 17) we obtain

$\mathrm{vec}(\mathbb{K}_{\mathcal{C}}^T) = (\mathbb{Z}_{\mathcal{C}} \otimes \mathbb{Z}_{\mathcal{C}})\, \mathrm{vec}(\mathbb{G}_{\mathcal{C}}^T) = \mathrm{vec}(\mathbb{Z}_{\mathcal{C}}\mathbb{G}_{\mathcal{C}}^T\mathbb{Z}_{\mathcal{C}}^T)$

so that $\mathbb{K}_{\mathcal{C}}^T = \mathbb{Z}_{\mathcal{C}}\mathbb{G}_{\mathcal{C}}^T\mathbb{Z}_{\mathcal{C}}^T$ and, therefore, $\mathbb{K}_{\mathcal{C}} = \mathbb{Z}_{\mathcal{C}}\mathbb{G}_{\mathcal{C}}\mathbb{Z}_{\mathcal{C}}^T$. The equivalence of the remaining three identities can be shown in the same way. $\square$

4

Undirected Graph Models

In a graphical model the interest is for the conditional independence structure of the variables, and the distinguishing feature of these models is that the collection of independence relationships characterizing the model is encoded by a graph. The first class of graphical models we consider is that of models associated with undirected graphs, which are usually referred to as *undirected graphical models* or *Markov random fields*. In the forthcoming chapters we will present the theory of models for bidirected graphs, known as bidirected graph models, and of models for directed acyclic and regression graphs, called directed acyclic graph and regression graph models, respectively. Accordingly, in order to keep the terminology consistent through this text we will slightly modify the traditional terminology and refer to models for undirected graphs as *undirected graph models*.

Discrete models for undirected graphs are among the very first classes of graphical models introduced in the literature. They have close connections with log-linear models and with systems investigated in other scientific areas such as statistical physics. A description of the origins of these models can be found in Cox and Wermuth (1996, section 2.13) and Lauritzen (1996, chapter 1). Here, we mention that the theory of undirected graph models presented in this chapter emerged from the seminal paper of Darroch *et al.* (1980), where the connection existing between log-linear models, conditional independence and undirected graphs was firstly established. Furthermore, we refer the reader to Edwards (2000, chapter 2) for a list of relevant classical monographs on this area and to Højsgaard *et al.* (2012) for an introduction to the available R packages (R Core Team, 2016) to deal with graphical models.

4.1 Graphs

A *graph* $G = (V, E)$ is a mathematical object that consists of a finite set of *vertices* V and a set of *edges* E, which are pairs of vertices. We write

$a \leftrightharpoons b$ to denote that the vertices $a, b \in V$ form an edge in G, and we say that a and b are *adjacent* or, alternatively, that they are *joined by an edge*. Conversely, we write $a \not\leftrightharpoons b$ for *non-adjacent* vertices. The graphs we consider are *simple*, i.e., they contain no multiple edges joining the same two vertices and no degenerate edge joining a vertex to itself. Furthermore, we only need three different types of edges that are typically labelled as *undirected, directed* and *bidirected*. A basic feature of graphs is that they are visual objects, and the usual way to picture a graph is drawing a dot for each vertex and joining two dots by a line $(-)$ for undirected edges, an arrow $(\rightarrow)$ for directed edges and an arrow with two heads $(\leftrightarrow)$ for bidirected edges.

Undirected Graphs

This chapter is devoted to graphical models for *undirected graphs*; i.e., graphs with only undirected edges, and when we want to highlight that a graph is undirected we write its edge set as $E^\sim$. For example, Figure 4.1 gives a graphical representation of the undirected graph $G = (V, E^\sim)$ with $V = \{a, b, c, d, e, f\}$ and

$$E^\sim = \{\{a,b\}, \{a,c\}, \{b,d\}, \{c,d\}, \{c,e\}, \{c,f\}, \{d,f\}, \{e,f\}\}.$$

We write $a \sim b$ to denote that a and b are joined by an undirected edge so that in this graph we have, for instance, $c \sim f$ because $\{c,f\} \in E^\sim$, but $a \not\sim f$ because $\{a,f\} \notin E^\sim$.

A *path* of length k from a to b is a sequence $a = v_0, \ldots, v_k = b$ of distinct vertices such that the vertices v_{r-1} and v_r are joined by an edge for all $r = 1, \ldots, k$. A *cycle* is a path with the modification that it begins and ends in the same vertex, i.e., $a = b$. For example, in the graph of Figure 4.1 the sequence (a, b, d, c, f) is a path from a to f, whereas (a, b, d, f, c, a) is a cycle with five vertices. A subset $C \subseteq V$ is said to *separate a from b* in G if all paths from a to b intersect C. The subset C is said to separate the subsets $A, B \subseteq V$ if it separates every $a \in A$ from every $b \in B$. Hence, both the sets $C_1 = \{c, d\}$ and $C_2 = \{b, c\}$ separate a from f in the graph of Figure 4.1 and, moreover, the same sets also separate $A = \{a\}$ form $B = \{e, f\}$.

If a and b are joined by an undirected edge we say that they are *neighbors*. The set of neighbors of a vertex a in G is denoted as $\mathrm{nb}_G(a)$, and we will omit the subscript when clear from the context which graph we are referring to. The set of neighbors of a subset $A \subseteq V$, $\mathrm{nb}(A)$, is the

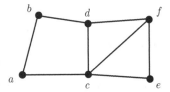

Figure 4.1 Example of undirected graph.

collection of the neighbors of the vertices in A that are not themselves elements of A; formally, $\mathrm{nb}(A) = (\cup_{a\in A}\mathrm{nb}(a))\backslash A$. Hence, in the graph of Figure 4.1 the neighbors of a are $\mathrm{nb}(a) = \{b,c\}$, whereas $\mathrm{nb}(b) = \{a,d\}$ so that if $A = \{a,b\}$ then $\mathrm{nb}(A) = \{c,d\}$.

4.2 Markov Properties for Undirected Graphs

A graphical model uses a graph to represent conditional independence relations holding among a set of variables. The vertices of the graph are associated with the variables, and the rules that translate properties of the graph into conditional independence statements are called *Markov properties*. In *undirected graph models* the variables of a random vector Y_V are indexed by the vertices of an undirected graph $G = (V, E^\sim)$, and in this context there are three Markov properties named pairwise, local and global.

The Pairwise Markov Property

The probability distribution of a random vector Y_V is said to obey the *pairwise Markov property* with respect to $G = (V, E^\sim)$ if for every $a, b \in V$ such that $\{a, b\} \notin E^\sim$ it holds that

$$Y_a \perp\!\!\!\perp Y_b | Y_{V\backslash\{a,b\}}.$$

For example, two conditional independence relations implied by the pairwise Markov property for the graph of Figure 4.1 are $Y_a \perp\!\!\!\perp Y_f | Y_{\{b,c,d,e\}}$ and $Y_b \perp\!\!\!\perp Y_c | Y_{\{a,d,e,f\}}$.

The Local Markov Property

The probability distribution of a random vector Y_V is said to obey the *local Markov property* with respect to $G = (V, E^\sim)$ if for every $a \in V$ it holds that

$Y_a \perp\!\!\!\perp Y_{V\setminus(\mathrm{nb}(a)\cup\{a\})}|Y_{\mathrm{nb}(a)}.$

For example, if Y_V is locally Markov with respect to the graph of Figure 4.1 then it holds that $Y_a \perp\!\!\!\perp Y_{\{d,e,f\}}|Y_{\{b,c\}}$, $Y_d \perp\!\!\!\perp Y_{\{a,e\}}|Y_{\{b,c,f\}}$ and $Y_c \perp\!\!\!\perp Y_b|Y_{\{a,d,e,f\}}.$

The Global Markov Property

The probability distribution of a random vector Y_V is said to obey the *global Markov property* with respect to $G = (V, E^\sim)$ if for any triple of pairwise disjoint subsets $A, B, S \subseteq V$ such that S separates A from B in G it holds that

$Y_A \perp\!\!\!\perp Y_B|Y_S.$

Examples of conditional independencies implied by the global Markov property for the graph of Figure 4.1 are $Y_a \perp\!\!\!\perp Y_{\{e,f\}}|Y_{\{c,d\}}$ and $Y_{\{b,d\}} \perp\!\!\!\perp Y_e|Y_{\{c,f\}}.$

Clearly, the global Markov property implies the local Markov property because $\mathrm{nb}(a)$ separates a from $V\setminus(\mathrm{nb}(a)\cup\{a\})$ for every $a \in V$ and, furthermore, it also implies the pairwise Markov property because $V\setminus\{a,b\}$ separates a from b whenever the two vertices are not joined by an edge. In addition, it can be shown that the local Markov property implies the pairwise Markov property (see Lauritzen, 1996, proposition 3.4) and therefore we can write

(global) $\Rightarrow$ (local) $\Rightarrow$ (pairwise).

The reverse implications are not true in general; however, a sufficient condition for the reverse implications to hold is that the joint density of Y_V with respect to a product measure is positive and continuous (Pearl and Paz, 1987); see also Lauritzen (1996, proposition 3.1 and theorem 3.7). It follows that in the case considered in this text, i.e., categorical variables with positive probability, the three Markov properties are equivalent,

(global) $\Leftrightarrow$ (local) $\Leftrightarrow$ (pairwise),

and in the following we will simply say that a distribution is Markov with respect to an undirected graph to mean that it satisfies all the three (undirected) Markov properties. It is also worth mentioning that, when Y

admits a density with respect to a product measure, an important role is played by a relevant factorization of the density with respect to the graph (see Lauritzen, 1996, section 3.2). Such factorization implies each of the three Markov properties for undirected graphs, and in the case where the density is positive and continuous also the reverse implication holds true. This fundamental result is known as the Hammersley–Clifford Theorem (Hammersley and Clifford, 1971); see also Lauritzen (1996, theorem 3.9).

We have thus seen that, among the three Markov properties, the global Markov property provides the most detailed information on the independence structure of the random vector because it gives a general criterion for assessing if two subvectors Y_A and Y_B are conditionally independent given a third subvector Y_S. More importantly, it can be shown that the global Markov property is *complete*; that is, the conditional independencies encoded by any undirected graph G under the global Markov property are the only conditional independencies that are simultaneously satisfied by all the distributions which are globally Markov with respect to G; see Geiger and Pearl (1993) and Studený (2005, p. 45) for a more detailed account on this topic. However, to practically deal with probability distributions satisfying this property, it is also useful to notice that the global Markov property is characterized by a smaller set of independence relationships, namely those with separating sets of the form $S = V \backslash (A \cup B)$. This is formally proved below.

Proposition 4.1 *The probability distribution of Y_V obeys the global Markov property with respect to $G = (V, E^\sim)$ if and only if for all pairs of nonempty disjoint subsets $A, B \subseteq V$ it holds that $Y_A \perp\!\!\!\perp Y_B | Y_{V \backslash (A \cup B)}$ whenever $V \backslash (A \cup B)$ separates A from B in G.*

Proof. If the distribution of Y_V is globally Markov with respect to G then every separation implies a conditional independence relationship and therefore it also holds that $Y_A \perp\!\!\!\perp Y_B | Y_{V \backslash (A \cup B)}$ whenever $V \backslash (A \cup B)$ separates A from B in G. Hence, it is sufficient to show the reverse implication; i.e., that the collection of independence relationships of the form $Y_A \perp\!\!\!\perp Y_B | Y_{V \backslash (A \cup B)}$ imply the global Markov property. In this case, if A, B and S is a triple of pairwise disjoint subsets of V such that S separates $A \neq \emptyset$ from $B \neq \emptyset$, then one can always find two disjoint sets D_1 and D_2 such that $D_1 \cup D_2 = V \backslash (A \cup B \cup S)$ and S separates $A \cup D_1$ from $B \cup D_2$. Hence, it holds that $Y_{A \cup D_1} \perp\!\!\!\perp Y_{B \cup D_2} | Y_S$ because

$S = V \backslash (A \cup D_1 \cup B \cup D_2)$ and it follows from the properties of conditional independence (see Lauritzen, 1996, section 3.1) that $Y_A \perp\!\!\!\perp Y_B | Y_S$. $\qquad \qquad \square$

4.3 The Log-linear Parameterization

If Y is a random vector with state space $\mathcal{I}$, a *statistical model* for Y is a family of probability distributions on $\mathcal{I}$. Our interest is in models characterized by collections of independence relationships, and a basic prerequisite to implement, in an efficient way, the restrictions defining the model is the specification of a suitable parameterization. A straightforward way to parameterize the probability distribution of a set of categorical variables is by means of their joint probability mass function, and we say that a model is a *saturated model* when the cell probabilities have no restrictions, apart from the sum-to-one constraint. Probabilities are easy to interpret but have the drawback that submodels of interest typically involve nonlinear constraints on these parameters. For instance, pairwise conditional independence relationships can be specified by requiring certain factorizations of the cell probabilities. For this reason, it is useful to develop alternative parameterizations such that submodels of interest correspond to linear subspaces of the parameter space of the saturated model.

An important advance in the analysis of categorical variables was to write the probability mass function as a log-linear expansion of the cell probabilities; see Agresti (2013, chapter 9) and Whittaker (1990, chapter 7). The rest of this section is devoted to the log-linear parameterization and shows how it can be used to implement conditional independence constraints.

The Binary Case

Our approach to the development of the log-linear parameterization heavily relies upon the connection existing between cross-product ratios and conditional independence relationships, described in Section 2.3, and on the theory of Möbius inversion given in Chapter 3. In order to readily exploit the results on Möbius inversion, it is useful to represent the probability mass function π^V of a binary vector Y_V as a vector indexed by the subsets of V. More concretely, we set $\pi^V = (\pi_D^V)_{D \subseteq V}$ where

$$\pi_D^V = p(Y_D = 1, Y_{V \setminus D} = 0) \quad \text{for every } D \subseteq V.$$

Thus we shall in the following switch between the case where π^V has the form of a probability table and the case where π^V is a vector indexed by the subsets of V. The former notation is perhaps easier to read, whereas the latter allows a compact representation of the results and a straightforward generalization to categorical variables with arbitrary number of levels.

We will refer to π^V as the *probability parameter* of Y_V, and the *log-linear parameter* λ^V of Y_V is defined as

$$\lambda^V = \mathbb{M}^T \log \pi^V, \tag{4.1}$$

where the logarithm is taken entrywise. Recall that we may omit the superscript when it is equal to V so that $\pi = \pi^V$ and $\lambda = \lambda^V$.

The quantity $\mathbb{M}$ in (4.1) is the Möbius matrix of Section 3.1.2, and it follows from Theorem 3.2 that λ constitutes a valid parameterization of the distribution because there exists a one-to-one relationship (a smooth bijection) between λ and π such that $\pi = \exp(\mathbb{Z}^T \lambda)$, where the exponential is also taken entrywise. Notice that the definition of the vector π, and therefore of the log-linear parameters in (4.1), requires the choice of a reference level for every variable, that in our case is the level "0". This is equivalent to specifying a cell to be set as baseline, and when the reference level of each variables is the value 0 the baseline cell is that indexed by the empty set. As a consequence, the value taken by λ is not invariant under relabeling of the two levels that are taken by the variables. However, such relabeling does not change the log-linear models we consider here. The choice of a reference cell in the specification of log-linear parameters is known as the *baseline* or *corner* constraint approach, whereas a common alternative approach is based on *symmetrical* constraints (see Agresti, 2013, chapter 9).

It follows from (3.3) that the entrywise version of the equality $\log \pi = \mathbb{Z}^T \lambda$ is given by

$$\log \pi_D = \sum_{D' \subseteq D} \lambda_{D'} \quad \text{for every } D \subseteq V,$$

where the individual log-linear parameters can be computed as

$$\lambda_D = \sum_{D' \subseteq D} (-1)^{|D \setminus D'|} \log \pi_{D'} \quad \text{for every } D \subseteq V. \tag{4.2}$$

If D is a singleton set, then λ_D is called a *main effect*, whereas for $|D|>1$ the parameter λ_D is called a $|D|$-*way interaction* or, equivalently, an *interaction of order* $(|D| - 1)$.

The interpretation of the value taken by a log-linear term is not straightforward, especially for high-order interactions. To this aim, it is useful to describe the entries of λ when V is a set of low cardinality such as $V = \{a, b, c\}$. For $D = \varnothing$ the term λ_D is the log-probability of the baseline cell of the table,

$$\lambda_\varnothing = \log \pi_\varnothing = \log p(Y_a = 0, Y_b = 0, Y_c = 0).$$

The main effect for $D = \{a\}$ is given by

$$\lambda_a = \log \frac{\pi_a}{\pi_\varnothing} = \log \frac{p(Y_a = 1, Y_b = 0, Y_c = 0)}{p(Y_a = 0, Y_b = 0, Y_c = 0)}.$$

The two-way interaction relative to $D = \{a, b\}$ is the logarithm of the odds ratio of the conditional distribution $Y_{\{a,b\}}|\{Y_c = 0\}$, introduced in Section 2.3,

$$
\begin{aligned}
\lambda_{\{a,b\}} &= \log \frac{\pi_{\{a,b\}}\pi_\varnothing}{\pi_a \pi_b} \\
&= \log \frac{p(Y_a = 1, Y_b = 1, Y_c = 0)p(Y_a = 0, Y_b = 0, Y_c = 0)}{p(Y_a = 1, Y_b = 0, Y_c = 0)p(Y_a = 0, Y_b = 1, Y_c = 0)} \\
&= \log \frac{p(Y_a = 1, Y_b = 1|Y_c = 0)p(Y_a = 0, Y_b = 0|Y_c = 0)}{p(Y_a = 1, Y_b = 0|Y_c = 0)p(Y_a = 0, Y_b = 1|Y_c = 0)} \\
&= \log \mathrm{cpr}_{cc}(a, a : b, b| \varnothing, c),
\end{aligned}
$$

and, therefore, $\lambda_{\{a,b\}} = 0$ is equivalent to the conditional independence relationship $Y_a \perp\!\!\!\perp Y_b|\{Y_c = 0\}$. Finally, the three-way interaction for $D = \{a, b, c\}$ involves the odds ratios of $Y_{\{a,b\}}|\{Y_c = 0\}$ and of $Y_{\{a,b\}}|\{Y_c = 1\}$ as follows,

$$
\begin{aligned}
\lambda_{\{a,b,c\}} &= \log \frac{\pi_{\{a,b,c\}} \pi_a \pi_b \pi_c}{\pi_{\{a,b\}} \pi_{\{a,c\}} \pi_{\{b,c\}} \pi_\varnothing} \\
&= \log \frac{p(Y_a = 1, Y_b = 1|Y_c = 1)p(Y_a = 0, Y_b = 0|Y_c = 1)}{p(Y_a = 1, Y_b = 0|Y_c = 1)p(Y_a = 0, Y_b = 1|Y_c = 1)} \\
&\quad -\log \frac{p(Y_a = 1, Y_b = 1|Y_c = 0)p(Y_a = 0, Y_b = 0|Y_c = 0)}{p(Y_a = 1, Y_b = 0|Y_c = 0)p(Y_a = 0, Y_b = 1|Y_c = 0)} \\
&= \log \mathrm{cpr}_{cc}(a, a : b, b|c, c) - \log \mathrm{cpr}_{cc}(a, a : b, b| \varnothing, c).
\end{aligned}
$$

Hence, $\lambda_{\{a,b,c\}} = \lambda_{\{a,b\}} = 0$ is equivalent $\mathrm{cpr}_{cc}(a, a : b, b|c, c) = \mathrm{cpr}_{cc}(a, a : b, b|\varnothing, c) = 1$ and, therefore, to $Y_a \perp\!\!\!\perp Y_b|Y_c$ by Corollary 2.4.

We now formally establish the connection between conditional independence relationships and the vanishing of log-linear parameters. In our approach, this is obtained as an immediate consequence of the joint application of Corollary 2.4 and Lemma 3.3. More concretely, Corollary 2.4 states the equivalence of certain conditional independence relationships and the fact that certain cross-product ratios are equal to one. On the other hand, Lemma 3.3 establishes the equivalence of certain vanishing entries of λ and the fact that certain contrasts of log-probabilities are equal to zero. The connection between the two results follows because the latter contrasts of log-probabilities and (the logarithm of) the relevant cross-product ratios are exactly the same quantities (see also La Rocca and Roverato, 2017).

Theorem 4.2   *For a vector Y_V of binary variables with probability parameter $\pi > 0$ let $\lambda = \mathbb{M}^T \log \pi$. Then, for a pair of disjoint nonempty subsets A and B of V the following conditions are equivalent:*

(i) $Y_A \perp\!\!\!\perp Y_B|Y_{V\setminus(A\cup B)}$;
(ii) *for every $D \subseteq V$ such that both $D \cap A \neq \varnothing$ and $D \cap B \neq \varnothing$ it holds that*

$$\lambda_D = 0;$$

(iii) *for every $A' \subseteq A$, $B' \subseteq B$, such that $A', B' \neq \varnothing$, and $C' \subseteq V\setminus(A\cup B)$ it holds that*

$$\log \pi_{A'\cup B'\cup C'} - \log \pi_{A'\cup C'} - \log \pi_{B'\cup C'} + \log \pi_{C'} = 0. \qquad (4.3)$$

Proof.   The equivalence of (i) and (iii) follows from Corollary 2.4 because the identity (4.3) is equal to $\log \mathrm{cpr}_{cc}(A', A : B', B|C', C) = 0$ with $C = V\setminus(A\cup B)$. The equivalence of (ii) and (iii) is that stated in Lemma 3.3 when $\theta = \lambda$ and $\omega = \log \pi$. $\square$

The Non-binary Case

In this section we extend the results given in the binary case to the general case of categorical variables with an arbitrary number of levels.

Let Y be a vector of categorical variables with state space $\mathcal{I}$ and restricted state space $\mathcal{J}$. We now define the probability parameter and the log-linear parameter of Y as a collection of vectors for $j \in \mathcal{J}$. More concretely, we represent $\mathcal{I}$ by means of a collection of binary tables $\{(j_D, 0_{V \setminus D})\}_{D \subseteq V}$ for all $j \in \mathcal{J}$ as in (2.3) and, accordingly, the probability mass function π of Y can be written as the collection of vectors

$$\pi_j = (\pi_{j_D})_{D \subseteq V} \text{ for all } j \in \mathcal{J}, \tag{4.4}$$

where

$$\pi_{j_D} = p(Y_D = j_D, Y_{V \setminus D} = 0).$$

Consequently, we can define the log-linear parameter λ of Y as the collection of vectors

$$\lambda_j = (\lambda_{j_D})_{D \subseteq V} \text{ for all } j \in \mathcal{J}, \tag{4.5}$$

where

$$\lambda_j = M^T \log \pi_j \text{ and } \lambda_{j_D} = (\lambda_j)_D,$$

with the convention that $j_\emptyset = \mathcal{J}_\emptyset = \emptyset$. The representation (4.4) is useful in practice because most properties that hold in the binary case can be readily generalized to the non-binary case by their iterative application to the vectors π_j for $j \in \mathcal{J}$. On the other hand, it is not an efficient representation of the probability mass function of Y because every pair of vectors π_j and $\pi_{j'}$ has nonempty intersection, in the sense that they have common entries. Indeed, for every $j, j' \in \mathcal{J}$ and $D \subseteq V$ such that $j_D = j'_D$ it holds that $\pi_{j_D} = \pi_{j'_D}$. Hence, the probability parameter of Y can be more efficiently written as $\pi = (\pi_{j_D})_{D \subseteq V, j_D \in \mathcal{J}_D}$. Similarly, we can write the log-linear parameter of Y as $(\lambda_{j_D})_{D \subseteq V, j_D \in \mathcal{J}_D}$.

We can now give the extension of Theorem 4.2 to the general case of categorical variables with arbitrary number of levels.

Corollary 4.3 *Let Y_V be a vector of categorical variables with probability parameter $\pi = (\pi_{j_D})_{D \subseteq V, j_D \in \mathcal{J}_D}$, $\pi > 0$, and log-linear parameter $\lambda = (\lambda_{j_D})_{D \subseteq V, j_D \in \mathcal{J}_D}$. Then, for a pair of disjoint nonempty subsets A and B of V the following conditions are equivalent:*

(i) $Y_A \perp\!\!\!\perp Y_B | Y_{V \setminus (A \cup B)}$;
(ii) *for every $D \subseteq V$ such that both $D \cap A \neq \emptyset$ and $D \cap B \neq \emptyset$ it holds that*

$$\lambda_{j_D} = 0,$$

for every $j \in \mathcal{J}$;
(iii) *for every $A' \subseteq A$, $B' \subseteq B$, such that $A', B' \neq \emptyset$, and $C' \subseteq V \backslash (A \cup B)$*
it holds that

$$\log \pi_{j_{A' \cup B' \cup C'}} + \log \pi_{j_{A' \cup C'}} + \log \pi_{j_{B' \cup C'}} - \log \pi_{j_{C'}} = 0, \qquad (4.6)$$

for every $j \in \mathcal{J}$.

Proof. The equivalence of (i) and (iii) follows from Corollary 2.8 because the identity (4.6) is equal to $\log \mathrm{cpr}_{cc}(j_{A'}, A : j_{B'}, B | j_{C'}, C) = 0$ with $C = V \backslash (A \cup B)$. The equivalence of (ii) and (iii) holds for every $j \in \mathcal{J}$ and is that stated in Lemma 3.3 when $\theta = \lambda_j$ and $\omega = \log \pi_j$. $\square$

4.4 Hierarchical Log-linear Models

The importance of the log-linear expansion of the cell probabilities rests on the fact that many interesting hypotheses can be specified as linear constraints on the log-linear parameters. Our interest is for models defined by the collection of conditional independence relationships encoded by an undirected graph. It follows directly from the definition of the pairwise and the local Markov properties, and from Proposition 4.1 for the global Markov property, that the conditional independencies that characterize each of the three Markov properties for undirected graphs are all of the form $Y_A \perp\!\!\!\perp Y_B | Y_{V \backslash (A \cup B)}$. This observation, together with the results stated in Theorem 4.2 and Corollary 4.3, shows that we can focus on log-linear models defined by vanishing collections of log-linear terms.

The Binary Case

The log-linear expansion of cell probabilities for the a binary random vector Y is given by

$$\log \pi_D = \sum_{D' \subseteq D} \lambda_{D'} \quad \text{for every } D \subseteq V,$$

and in the saturated model no restriction is posed on the log-linear parameters where the only fixed term is $\lambda_\emptyset$, resulting from the sum-to-one constraint of probabilities. It also deserves noticing that, apart from $\lambda_\emptyset$, the entries of λ are *variation independent*, in the sense that setting some parameters to particular values does not restrict the valid range of other parameters.

When the interest is for log-linear models defined by zero constraints, a relevant family of models is that of *hierarchical log-linear models*. A log-linear model, defined by zero restrictions on the log-linear terms, is called hierarchical if whenever an interaction term is fixed to zero then all higher-order interaction terms involving the same variables are also fixed to zero. More formally, this means that in a hierarchical log-linear model, for every $D \subseteq V$ with $D \neq \emptyset$

the restriction $\lambda_D = 0$ implies the restriction $\lambda_E = 0$ for all $E \supseteq D$

or, equivalently,

λ_D unrestricted implies $\lambda_{D'}$ unrestricted for all $D' \subseteq D$ with $D' \neq \emptyset$.

Nonhierarchical models are rarely used in practice because, in most applications, it is not meaningful to include higher-order interaction terms without incorporating the lower-order interactions composed from the variables. Furthermore, the family of hierarchical log-linear models is sufficiently general to encompass all the models defined by the conditional independencies encoded by an undirected graph, under any of the three Markov properties. This can be seen by noticing that the constraints on the log-linear terms given in point (ii) of Theorem 4.2 are of the hierarchical type. As an example, we consider the case $V = \{a, b, c\}$ so that λ is made up of as few as eight terms that are all present in the linear expansion of $\log \pi_V$,

$$\log \pi_{\{a,b,c\}} = \lambda_\emptyset + \lambda_a + \lambda_b + \lambda_c + \lambda_{\{a,b\}} + \lambda_{\{a,c\}} + \lambda_{\{b,c\}} + \lambda_{\{a,b,c\}}.$$
(4.7)

Hence, the hierarchical model with no second-order interaction, i.e., $\lambda_{\{a,b,c\}} = 0$, is such that,

$$\log \pi_{\{a,b,c\}} = \lambda_\emptyset + \lambda_a + \lambda_b + \lambda_c + \lambda_{\{a,b\}} + \lambda_{\{a,c\}} + \lambda_{\{b,c\}},$$
(4.8)

whereas in the hierarchical model with one first-order interaction set to zero, say $\lambda_{\{a,b\}} = 0$, also $\lambda_{\{a,b,c\}}$ must be fixed to zero,

$$\log \pi_{\{a,b,c\}} = \lambda_\emptyset + \lambda_a + \lambda_b + \lambda_c + \lambda_{\{a,c\}} + \lambda_{\{b,c\}}.$$
(4.9)

There exists a close connection between simplicial complexes of subsets of V and hierarchical log-linear model. For any hierarchical model the collection of sets indexing its non-vanishing terms forms a simplicial complex $\mathcal{C}$ and, conversely, any simplicial complex $\mathcal{C}$ of subsets of V can be used to define a hierarchical model. However, the most parsimonious way to specify a hierarchical model is by means of the collection of maximal sets of $\mathcal{C}$, with respect to inclusion, that is called the *generating class* of the model, and denoted by $\mathcal{C}^\uparrow$. Accordingly, we call $\mathcal{C}$ the *expanded generating* class and denote the relative model by either $M(\mathcal{C})$ or $M(\mathcal{C}^\uparrow)$. Hence, for $V = \{a, b, c\}$ the generating class of the saturated model is $\mathcal{C}^\uparrow = \{\{a, b, c\}\}$ whereas the expanded generating class for the model in (4.8) is

$$\mathcal{C} = \{\emptyset, \{a\}, \{b\}, \{c\}, \{a, b\}, \{a, c\}, \{b, c\}\},$$

and therefore $\mathcal{C}^\uparrow = \{\{a, b\}, \{a, c\}, \{b, c\}\}$. For the model in (4.9) it holds that $\mathcal{C}^\uparrow = \{\{a, c\}, \{b, c\}\}$. Furthermore, for $\mathcal{C}^\uparrow = \{\{a\}, \{b, c\}\}$ we have

$$\log \pi_{\{a,b,c\}} = \lambda_\emptyset + \lambda_a + \lambda_b + \lambda_c + \lambda_{\{b,c\}}, \tag{4.10}$$

whereas the generating class $\{\{a\}, \{b\}, \{c\}\}$ gives the *main effects model* with

$$\log \pi_{\{a,b,c\}} = \lambda_\emptyset + \lambda_a + \lambda_b + \lambda_c. \tag{4.11}$$

We now turn to the use of undirected graphs for describing the independence structure of Y. By Theorem 4.2, a necessary and sufficient condition for the conditional independence of two variables given the remaining variables, say $Y_a \perp\!\!\!\perp Y_b | Y_{V \setminus \{a,b\}}$, is that $\lambda_D = 0$ whenever D is a superset of $\{a, b\}$. Hence, if $G = (V, E^\sim)$ is an undirected graph with vertex set V, and the log-linear parameter λ is such that for every pair $a, b \in V$ such that $a \neq b$, it holds that $\lambda_{\{a,b\} \cup D'} = 0$ for every $D' \subseteq V \setminus \{a, b\}$, then the distribution of Y is pairwise Markov with respect to G. Furthermore, when the distribution is positive, it also obeys both the local and the global Markov properties with respect to G.

We can therefore associate any hierarchical log-linear model with an undirected graph, which we call the *independence graph* of $M(\mathcal{C}^\uparrow)$. This is the graph G, with vertex set V, constructed by joining two vertices $a, b \in V$ by an edge if and only if $\lambda_{\{a,b\}}$ is unconstrained in $M(\mathcal{C}^\uparrow)$ or, equivalently, if and only if there exists a $C \in \mathcal{C}^\uparrow$ such that $\{a, b\} \subseteq C$. By

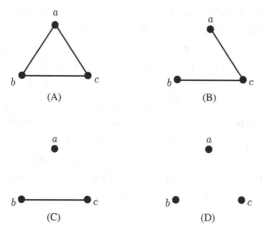

Figure 4.2 Four independence graphs for $Y_{\{a,b,c\}}$: under the global Markov property (A) encodes no independence relationship; (B) encodes the independence $Y_a \perp\!\!\!\perp Y_b | Y_c$; (C) encodes the independence $Y_a \perp\!\!\!\perp Y_{\{b,c\}}$; and (D) encodes the mutual independence $Y_a \perp\!\!\!\perp Y_b \perp\!\!\!\perp Y_c$.

construction, all the distributions in $M(\mathcal{C}^{\uparrow})$ obey the pairwise Markov property with respect to G, and also the local and global Markov properties in the case, considered here, of positive distributions. For instance, when $V = \{a, b, c\}$, the saturated model is associated with the graph with no missing edge of Figure 4.2A. The same independence graph is associated with the no-second-order interaction model in (4.8) because, although $\lambda_{\{a,b,c\}} = 0$, all first-order interactions are unconstrained. The model with $\lambda_{\{a,b\}} = \lambda_{\{a,b,c\}} = 0$ in (4.9) leads to the graph of Figure 4.2B where $a \not\sim b$ implies $Y_a \perp\!\!\!\perp Y_b | Y_c$. The model with $\lambda_{\{a,b\}} = \lambda_{\{a,c\}} = \lambda_{\{a,b,c\}} = 0$ in (4.10) leads to the graph of Figure 4.2C where $a \not\sim b$ implies $Y_a \perp\!\!\!\perp Y_b | Y_c$ and $a \not\sim c$ implies $Y_a \perp\!\!\!\perp Y_c | Y_b$. Furthermore, the global Markov property also implies the independencies $Y_a \perp\!\!\!\perp Y_{\{b,c\}}$, $Y_a \perp\!\!\!\perp Y_b$ and $Y_a \perp\!\!\!\perp Y_c$. Finally, the independence graph corresponding to the model in (4.11) is that given in Figure 4.2D, which encodes all the pairwise conditional independence relationships and, under the global Markov property, the mutual independence of the three variables, $Y_a \perp\!\!\!\perp Y_b \perp\!\!\!\perp Y_c$.

We close this section by noticing that we can use the expanded generating class $\mathcal{C}$ of a hierarchical model to extract the subvector $\lambda_{\mathcal{C}}$

of λ. Furthermore, if we denote by $\mathcal{P}(V)$ the expanded generating class of the saturated model and set $\bar{\mathcal{C}} = \mathcal{P}(V) \backslash \mathcal{C}$ then we can write $\lambda = (\lambda_{\mathcal{C}}, \lambda_{\bar{\mathcal{C}}})$ where, under $M(\mathcal{C})$, $\lambda_{\bar{\mathcal{C}}} = 0$. Correspondingly, we can write $\pi = (\pi_{\mathcal{C}}, \pi_{\bar{\mathcal{C}}})$ and, if $\mathbb{Z}_{\mathcal{C}}$ is the submatrix of $\mathbb{Z}$ whose entries are indexed by $\mathcal{C} \times \mathcal{C}$ then it holds that $\log \pi_{\mathcal{C}} = \mathbb{Z}_{\mathcal{C}}^T \lambda_{\mathcal{C}}$ and thus, by Corollary 3.6, that $\lambda_{\mathcal{C}} = \mathbb{M}_{\mathcal{C}}^T \log \pi_{\mathcal{C}}$. This makes it clear that for a log-linear model with expanded generating class $\mathcal{C}$, the probability parameter π can be partitioned into $\pi = (\pi_{\mathcal{C}}, \pi_{\bar{\mathcal{C}}})$, where $\pi_{\bar{\mathcal{C}}}$ can be obtained as a function of $\lambda_{\mathcal{C}}$ and, therefore, of $\pi_{\mathcal{C}}$. Consequently, the cell probabilities identified by $\mathcal{C}$, i.e., $\pi_{\mathcal{C}}$, are sufficient to characterize the probability distribution of Y.

The Non-binary Case

The extension of the theory presented in the above subsection for the binary case to categorical variables with arbitrary number of levels is straightforward, and it is mainly a matter of notation.

In the general case where Y is a vector of polytomous variables the log-linear expansion of cell probabilities is

$$\log \pi_{j_D} = \sum_{D' \subseteq D} \lambda_{j_{D'}} \quad \text{for every } D \subseteq V \text{ and } j \in \mathcal{J},$$

and the constraint $\lambda_D = 0$ in the binary case needs to be reformulated as $\lambda_{j_D} = 0$ for every $j \in \mathcal{J}$. Then, the definition of hierarchical log-linear model is generalized accordingly, so that also in the non-binary case every hierarchical model is uniquely associated with an (extended) generating class. The connection between vanishing log-linear parameters and conditional independence follows from Corollary 4.3, which generalizes Theorem 4.2. This implies that, in the non-binary case, the independence graph associated with a hierarchical log-linear model can be constructed from the generating class of the model in the same was as in the binary case.

4.5 Log-linear Graphical Models

Every hierarchial log-linear model has an independence graph associated with it. However, different hierarchial models may have the same independence graph. For example, both the saturated model $\{\{a, b, c\}\}$

and the no-second-order interaction model with generating class $\{\{a,b\},\{a,c\},\{b,c\}\}$ are associated with the complete graph of Figure 4.2A. This is due to the fact that the two models differ for the constraint $\lambda_{\{a,b,c\}} = 0$ which does not implies any conditional independence. Therefore, an independence graph does not in general provide a complete description of the structure of a hierarchical model. The family of graphical log-linear models is the subclass of hierarchical models characterized by their independence graph (Darroch *et al.*, 1980; Edwards and Kreiner, 1983).

Definition 4.5.1 An *undirected graph model* with graph G is the hierarchical log-linear model whose generating class is the set of the cliques of G. Thus, the graph G defines the model, and we write $M(G)$ to denote the graphical model with independence graph G.

Undirected graph models are therefore models that can be interpreted solely in terms of conditional independence relationships, and the latter can be completely derived from the independence graph through graph separation. For example, the graphical model defined by the graph of Figure 4.1 is the hierarchical log-linear model with generating class $\{\{a,b\},\{a,c\},\{b,d\},\{c,d,f\},\{c,e,f\}\}$. In the binary case this model has 17 non-vanishing log-linear terms, including $\lambda_\varnothing$. The smallest hierarchical log-linear model with the same independence graph has generating class

$$\{\{a,b\},\{a,c\},\{b,d\},\{c,d\},\{c,e\},\{c,f\},\{d,f\},\{e,f\}\}.$$

This model is not graphical, and it can be obtained from the relevant graphical model by imposing the additional constraints $\lambda_{\{c,d,f\}} = \lambda_{\{c,e,f\}} = 0$. It has the advantage that it is more parsimonious than the graphical model with the same independence graph but, on the other hand, the additional constraints have no conditional independence interpretation.

4.6 Data, Estimation and Testing

Let $Y^{(1)} = y^{(1)},\ldots,Y^{(n)} = y^{(n)}$ be a random sample of n independent and identically distributed observations from a distribution with probability mass function π. Samples from a vector of categorical variables can be conveniently presented as a *contingency table* $\{n(Y=i)\}_{i\in\mathcal{I}}$

Table 4.1 Infant survival data: cross-classification of
survival of infants by amount of prenatal care and
clinic.

			Survival
Clinic	**Care**	**No**	**Yes**
Clinic A	Less	3	176
	More	4	293
Clinic B	Less	17	197
	More	2	23

where the sampling units are classified according to the levels of the
variables, i.e., the cells of the table. For instance, Table 4.1 gives the
contingency table of the *infant survival* data set provided by Bishop
et al. (1975) (see also Bishop *et al.*, 2007; Whittaker, 1990, section 1.2)
that gives information on the survival of 715 infants attending two
clinics and the amount of care received by the mother.

The entry $n(Y = i)$ is called the count of the cell i, and it is the number
of observations falling in the cell i, formally computed as
$n(Y = i) = \sum_{r=1}^{n} 1(Y^{(r)} = i)$; recall that $1(\cdot)$ is the indicator function.
It is also useful to consider tables of *marginal counts* that are tables
produced by cross-classifying the observations on the basis of a subset
of the variables. The table of marginal counts relative to a nonempty
subset $A \subseteq V$ is denoted by $\{n(Y_A = i_A)\}_{i_A \in \mathcal{I}_A}$ where $n(Y_A = i_A) =$
$\sum_{r=1}^{n} 1(Y_A^{(r)} = i_A)$. For instance, Table 4.2 gives the marginal contingency tables of the pairs of variables {Clinic, Survival} and {Care,
Clinic} for the infant survival data.

We remark that the marginal counts can also be obtained from a
larger contingency table by marginalizing over the variables not in A
so that $n(Y_A = i_A) = \sum_{i_{V \backslash A} \in \mathcal{I}_{V \backslash A}} n(Y_A = i_A, Y_{V \backslash A} = i_{V \backslash A})$; likewise,
$n = \sum_{i \in \mathcal{I}} n(Y = i)$.

We now consider the problem of computing the maximum likelihood
(ML) estimate of the parameters of a hierarchical log-linear model. To
keep the presentation at a simpler level, we assume that the cell counts
are all strictly positive and refer to Lauritzen (1996, chapter 4) and
Agresti (2013, section 10.6) for an extension of the results of this section
to the case of empty cells and sparse contingency tables.

Table 4.2 Infant survival data: marginal tables of counts for the pairs of variables {Clinic, Survival} and {Care, Clinic}.

	Survival			Care	
Clinic	**No**	**Yes**	**Clinic**	**Less**	**More**
Clinic A	7	469	Clinic A	179	297
Clinic B	19	220	Clinic B	214	25

Under the *multinomial sampling scheme* considered here, the sample size n is fixed and the observations are independent and identically distributed. Thus, the sampling distribution of counts follows a multinomial distribution with probability mass function

$$p\{n(Y = i)\} = \frac{n!}{\Pi_{i \in \mathcal{I}} n(Y = i)!} \prod_{i \in \mathcal{I}} p(Y = i)^{n(Y=i)} \text{ for all } i \in \mathcal{I}.$$

(4.12)

The *expected count* of the of the cell i, $E\{n(Y = i)\}$, is denoted by $m(Y = i)$ and it can be computed by multiplying by n the relevant cell probability,

$$m(Y = i) = np(Y = i) \quad \text{for all } i \in \mathcal{I}.$$

It follows that the ML estimate can be equivalently computed with respect to expected counts because

$$\hat{p}(Y = i) = \frac{\hat{m}(Y = i)}{n} \text{ for every } i \in \mathcal{I}.$$

Equation (4.12) leads to the log-likelihood function

$$l(\pi) = \sum_{i \in \mathcal{I}} n(Y = i) \log p(Y = i),$$

(4.13)

and it follows immediately from (4.13) that for the saturated model, where $\mathcal{C}^{\uparrow} = \{V\}$, the sufficient statistic is the table of observed counts $\{n(Y = i)\}_{i \in \mathcal{I}}$ and, furthermore, it can be shown that the observed counts are the ML estimates of the expected counts,

$$\{\hat{m}(Y = i)\}_{i \in \mathcal{I}} = \{n(Y = i)\}_{i \in \mathcal{I}}.$$

The rules to compute the ML estimates for hierarchical log-linear models where first derived by Birch (1963) and are usually referred to as the *Birch's results*. They state that for an arbitrary hierarchical log-linear model with generating class $C^\uparrow = \{C_1, \ldots C_k\}$ the minimal sufficient statistic is the set of the marginal tables of counts corresponding to the maximal interactions of the model, i.e.,

$$\{n(Y_C = i_C)\}_{i_C \in \mathcal{I}_C} \text{ for every } C \in C^\uparrow,$$

and the ML estimates of the marginal distributions relative to the maximal interactions of the models are directly obtained from the sufficient statistics as follows

$$\{\hat{m}(Y_C = i_C)\}_{i_C \in \mathcal{I}_C} = \{n(Y_C = i_C)\}_{i_C \in \mathcal{I}_C} \text{ for every } C \in C^\uparrow. \quad (4.14)$$

Hence, the ML estimate of $\{m(Y = i)\}_{i \in \mathcal{I}}$ must satisfy both equations (4.14) and the restrictions defining the model. This is formally stated below.

Proposition 4.4   *The maximum likelihood estimate of $\{m(Y = i)\}_{i \in \mathcal{I}}$, under the hierarchical log-linear model with generation class $C^\uparrow$, based on a random sample from the multinomial distribution, exists and it is the unique $\{\hat{m}(Y = i)\}_{i \in \mathcal{I}}$ that satisfies both the system of equation (4.14) and the restrictions defining $M(C^\uparrow)$.*

A proof of this result can be found, for instance, in Lauritzen (1996, theorem 4.8). Here, it deserves noticing that log-linear models are regular exponential families and the result of Proposition 4.4 follows from the general properties of this class of models, where the ML estimates can be found by equating the sufficient statistics to their expectation. Although Proposition 4.4 is stated assuming a multinomial sampling scheme, it can be shown that it is also valid under the *Poisson sampling scheme*. Moreover, in the case of *independent multinomial sampling*, where certain marginal distributions are fixed by the sampling design, Proposition 4.4 holds in the case where the fixed marginals correspond to interactions that are present in the model; see Agresti (2013) for further details.

Many log-linear models have no closed-form ML estimate, and the solutions to the likelihood equations are computed by means of iterative procedures. A simple algorithm commonly used in this context is called *iterative proportional fitting* (IPF), and it works by successively adjusting the ML estimates of the marginal tables corresponding to the sufficient statistics (Deming and Stephan, 1940; Darroch and Ratcliff, 1972). The IPF algorithm for computing the ML estimate of a hierarchical model with generating class $C^\uparrow = \{C_1, \ldots, C_k\}$ has the following steps:

1. Start with an initial estimate $\{\widetilde{m}_0(Y = i)\}_{i \in \mathcal{I}}$ having association structure not more complex than $M(C^\uparrow)$. A common choice is

   $\widetilde{m}_0(Y = i) = 1$ for all $i \in \mathcal{I}$.

2. For every $s = 1, \ldots, k$ set $C = C_s$ and apply the updating step

   $$\widetilde{m}_s(Y = i) = \widetilde{m}_{s-1}(Y = i) \times \frac{n(Y_C = i_C)}{\widetilde{m}_{s-1}(Y_C = i_C)} \text{ for every } i \in \mathcal{I}.$$

3. If the maximal difference between the sufficient statistics $\{n(Y_C = i_C)\}_{i_C \in \mathcal{I}_C}$ and their fitted values $\{\widetilde{m}_k(Y_C = i_C)\}_{i_C \in \mathcal{I}_C}$ is sufficiently close to zero for all $C \in C^\uparrow$ then set $\{\widehat{m}(Y = i)\}_{i \in \mathcal{I}} = \{\widetilde{m}_k(Y = i)\}_{i \in \mathcal{I}}$ and stop. Otherwise, set $\{\widetilde{m}_0(Y = i)\}_{i \in \mathcal{I}} = \{\widetilde{m}_k(Y = i)\}_{i \in \mathcal{I}}$ and go to the step 2 above.

For details on the convergence of the IPF algorithm to the ML estimate we refer to Whittaker (1990, section 4.7) and references therein. Here we remark that the IPF works even in the case when there exists an explicit expression for calculating the ML estimate, and in this case the algorithm has the nice property that it converges to the ML estimate after a finite number of cycles.

   As an example, we show the application of the IPF algorithm to the infant survival data. We consider the model with $C^\uparrow = \{\{\text{Clinic, Survival}\}, \{\text{Care, Clinic}\}\}$ corresponding to $Y_{\text{Care}} \perp\!\!\!\perp Y_{\text{Survival}} | Y_{\text{Clinic}}$ and therefore to the independence graph of Figure 4.2B where Clinic is c and Care and Survival are a and b, respectively. The tables below give the estimates of the expected counts and the implied marginal tables for the pairs {Clinic, Survival} and {Care, Clinic}, which should be compared with the marginal counts in Table 4.2 to monitor the convergence of the procedure. We set the initial values constant and equal to 1, thereby specifying an initial model with no interaction terms, and therefore with simpler structure than the model being fitted.

Clinic	Care	Survival	
		No	Yes
A	Less	1	1
	More	1	1
B	Less	1	1
	More	1	1

Clinic	Survival	
	No	Yes
A	2	2
B	2	2

Clinic	Care	
	Less	More
A	2	2
B	2	2

Next, we apply step 2 of the algorithm to adjust the margin $C = \{$Clinic, Survival$\}$, thereby obtaining the updated estimate,

Clinic	Care	Survival	
		No	Yes
A	Less	3.5	234.5
	More	3.5	234.5
B	Less	9.5	110.0
	More	9.5	110.0

Clinic	Survival	
	No	Yes
A	7	469
B	19	220

| | Care | |
Clinic	Less	More
A	238.0	238.0
B	119.5	119.5

Subsequently, we apply step 2 of the algorithm to adjust the margin $C = \{$Care, Clinic$\}$ and this operation gives the following estimate

| | | Survival | |
Clinic	Care	No	Yes
A	Less	2.63	176.37
	More	4.37	292.63
B	Less	17.01	196.99
	More	1.99	23.01

| | Survival | |
Clinic	No	Yes
A	7	469
B	19	220

| | Care | |
Clinic	Less	More
A	179	297
B	214	25

As both the $\{$Clinic, Survival$\}$-marginal and the $\{$Care, Clinic$\}$-marginal satisfy the likelihood equations the algorithm has converged, and there is no need for further iterations. The fitted expected counts can then be divided by $n = 715$ to obtain the ML estimates of the cell probabilities. Notice that convergence has required a single cycle because, as we will see later, this model admits an explicit expression for the ML estimate.

We now turn to model fitting. The basic statistic used in this context is the *deviance* which, for a model $M(\mathcal{C})$, is computed as twice the difference between the log-likelihood of the saturated model and the log-likelihood taken over $M(\mathcal{C})$,

$$\mathrm{dev}(\mathcal{C}) = 2\sum_{i \in \mathcal{I}} n(Y = i)\log\frac{n(Y = i)}{\hat{m}(Y = i)}.$$

For hierarchical log-linear models the deviance has the same form as the *likelihood-ratio statistic* for testing a log-linear model against the corresponding saturated model. When $M(\mathcal{C})$ holds, the deviance has asymptotic chi-squared distribution with degrees of freedom given by the number of constrains of $M(\mathcal{C})$, i.e., $2^{|V|} - |\mathcal{C}|$. For example, if for the infant survival data we consider the model with generation class $\mathcal{C}^\uparrow = \{\{\text{Clinic, Survival}\}, \{\text{Care, Clinic}\}\}$ we can use the ML estimate computed above to see that $\mathrm{dev}(\mathcal{C}) = 0.08$. Because the number of constraints of this model is equal to two, then it makes sense to compare the value of the deviance with the chi-squared quantile $\chi^2_{2;0.95} = 5.99$, thereby showing that this model provides a very good fit of the data (p-value $= 0.97$).

For two models M_0 and M_1 we write $M_0 \subseteq M_1$ if M_0 can be obtained from M_1 by imposing additional restrictions. It is easy to see that if M_0 and M_1 are hierarchical log-linear models with expanded generating classes $\mathcal{C}_0$ and $\mathcal{C}_1$, respectively, then $M_0 \subseteq M_1$ if and only if $\mathcal{C}_0 \subseteq \mathcal{C}_1$. The deviance statistic to compare the two models is the difference

$$\mathrm{dev}(\mathcal{C}_0|\mathcal{C}_1) = \mathrm{dev}(\mathcal{C}_0) - \mathrm{dev}(\mathcal{C}_1) = 2\sum_{i \in \mathcal{I}} n(Y = i)\log\frac{\hat{m}_1(Y = i)}{\hat{m}_0(Y = i)}.$$

$$(4.15)$$

When $M(\mathcal{C}_0)$ holds the deviance difference has asymptotic chi-squared distribution with degrees of freedom given by the number of additional constrains of $M(\mathcal{C}_0)$ compared to $M(\mathcal{C}_1)$, i.e., $|\mathcal{C}_1| - |\mathcal{C}_0|$.

4.7 Graph Decomposition and Decomposable Graphs

This section introduces some definitions and results from the theory of undirected graphs that can be used to identify relevant features of graphical models.

If $G = (V, E^\sim)$ is an undirected graph, the *subgraph* of G induced by the subset $A \subseteq V$, denoted by $G_A = (A, E_A)$, is the graph with vertex set A and all those edges which join two vertices that are both in A. An undirected graph $G = (V, E^\sim)$ is *complete* if all its vertices are mutually adjacent and we say that a subset $A \subseteq V$ is complete when G_A is complete. We denote by $\mathcal{V} = \mathcal{V}(G)$ the collection of all the complete subsets of V, and we remark that $\mathcal{V}$ is a simplicial complex. This can be seen by noticing that both all singleton subsets of V and the empty set are complete because they induce subgraphs which have exactly one vertex and no vertices, respectively. Accordingly, in the following we will refer to $\mathcal{V}$ as the *simplicial complex relative to* G. A subset $C \subseteq V$ is called a *clique* if it is maximally complete, i.e., C is complete, and if $C \subset D$, then D is not complete. For example, in the graph of Figure 4.1 the set $\{c, d\}$ is complete but it is not maximally complete because it is a subset of the complete set $\{c, d, f\}$. In parallel with the generating class of a hierarchical log-linear model that is formed by the maximal sets of a simplicial complex, also the cliques of G are the maximal sets of $\mathcal{V}$, and thus we denote the set of the cliques of G by $\mathcal{V}^\uparrow = \{C_1, \ldots, C_k\}$. An undirected graph can be identified by the set of its cliques, so that, for instance, the graph in Figure 4.1 is identified by $\mathcal{V}^\uparrow = \{\{a, b\}, \{a, c\}, \{b, d\}, \{c, d, f\}, \{c, e, f\}\}$.

Definition 4.7.1 The pair (A, B) of subsets of V such that $A \cup B = V$ forms a decomposition of $G = (V, E^\sim)$ if it holds that

(i) $A \cap B$ is a complete subset of V;
(ii) $A \cap B$ separates $A \setminus B$ from $B \setminus A$ in G.

The subgraphs G_A and G_B are two *components* of G. A decomposition is *proper* if both $A \setminus B \neq \emptyset$ and $B \setminus A \neq \emptyset$, and a graph is said to be *prime* if it does not admit any decomposition. For example, the pair $(\{a, b, c, d\}, \{c, d, e, f\})$ forms a decomposition of the graph in Figure 4.1. The components associated with this decomposition, given in Figure 4.3, are $G_{\{a,b,c,d\}}$, that is prime, and $G_{\{c,e,d,f\}}$, that is not prime because it admits the decomposition $(\{c, d, f\}, \{c, e, f\})$. Any undirected graph can be recursively decomposed into its *prime components* (Diestel, 1990). Figure 4.4 displays the prime components of the graph in Figure 4.1. In this example, the prime components $G_{\{c,d,f\}}$ and $G_{\{c,e,f\}}$ are complete subgraphs, and their vertex sets are cliques of G.

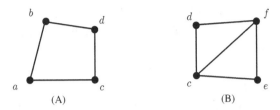

Figure 4.3 Components relative to the decomposition
$(\{a,b,c,d\}, \{c,d,e,f\})$ of the graph G in Figure 4.1: (A) $G_{\{a,b,c,d\}}$,
(B) $G_{\{c,d,e,f\}}$.

Figure 4.4 Prime components of the graph in Figure 4.1.

Conversely, the prime component $G_{\{a,b,c,d\}}$ is not complete, but it cannot
be further decomposed.

The recursive decomposition of a graph into its prime components
can be used to define the distinguished subclass of decomposable
graphs.

Definition 4.7.2   An undirected graph G is said to be *decomposable* if
either (i) it is complete, or (ii) there exists a proper decomposition (A, B)
of G such that both G_A and G_B are decomposable.

This definition is recursive, but it is well posed because the decompositions are required to be proper. To check whether a graph is decomposable it is sufficient to recursively decompose it into its prime
components, and the graph is decomposable if the prime components
are all complete, otherwise it is *nondecomposable*. It follows that a
graph is decomposable if and only if the vertex sets of its prime
components are the cliques of the graph. Thus, the graph in Figure 4.1
is nondecomposable, whereas it is not difficult to check that the graph in
Figure 4.5 is decomposable.

There exist a number of alternative ways to characterize the class of
decomposable graphs (see Cowell *et al.*, 1999, chapter 4). Every alternative characterization identifies a useful property of decomposable

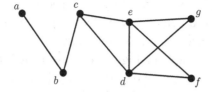

Figure 4.5 Example of decomposable graph.

graphs. For example, the fact that the class of decomposable graphs coincides with the class of chordal graphs provides a useful way to visually check decomposability in small graphs. A *chord* is a pair of non-consecutive vertices in a cycle that are joined by an edge. A cycle is *chordless* if it has no chords, and an undirected graph is said to be *chordal* or *triangulated* if it has no chordless cycle with four or more vertices. For example, in the graph of Figure 4.1 the cycle (c, d, f, e, c) possesses the cord $c \sim f$, whereas (a, b, d, c, a) is a chordless cycle with four vertices. On the other hand, the graph in Figure 4.5 has no chordless cycles. It can be shown that an undirected graph is chordal if and only if it is decomposable (see Cowell et al., 1999, theorem 4.4). A visual inspection of the graphs in Figures 4.5 and 4.1 allows one immediately to asses that the former is triangulated, and therefore decomposable, whereas the latter is not.

Let $(C_1, \ldots, C_k)$ be a sequence of complete subsets of the vertex set V. For all $r = 2, \ldots, k$ let

$$S_r = (C_1 \cup \ldots \cup C_{r-1}) \cap C_r.$$

The sequence $(C_1, \ldots, C_k)$ is said to be perfect if for every $r = 2, \ldots, k$ there is an $s < r$ such that $S_r \subseteq C_s$. This condition is known as the *running intersection* property, and the proof of the following result can be found in Lauritzen (1996, proposition 2.17).

Proposition 4.5 *An undirected graph is decomposable if and only if its cliques can be numbered to form a perfect sequence.*

For instance, the sequence $(\{a, b\}, \{b, c\}, \{c, d, e\}, \{d, e, f\}, \{d, e, g\})$ is a perfect sequence of the cliques of the graph in Figure 4.5. It is easy to see that any decomposable graph can be recursively decomposed by starting from a perfect sequence $(C_1, \ldots, C_k)$ of its cliques, and then by decomposing $G_{C_1 \cup \ldots \cup C_r}$ into the components $G_{C_1 \cup \ldots \cup C_{r-1}}$ and G_{C_r} for every $r = k, (k - 1), \ldots, 2$. Every step of this procedure is a decomposition and has a separating set associated with it, which we denote by S_r. Accordingly, every perfect sequence of cliques is associated with a sequence $(S_2, \ldots, S_k)$ of *separators* of G. The sequence of separators, for the graph in Figure 4.5, associated with the perfect sequence of cliques given above is $(\{b\}, \{c\}, \{d, e\}, \{d, e\})$. As this example shows, separators need not be distinct. Furthermore, apart from their ordering, separators are the same for all perfect sequences of cliques. For example, another perfect sequence of the cliques of the same graph is $(\{e, d, f\}, \{d, e, g\}, \{c, d, e\}, \{b, c\}, \{a, b\})$ and the associated sequence of separators is $(\{d, e\}, \{d, e\}, \{c\}, \{b\})$.

Formal algorithms have been developed for deciding whether an undirected graph is chordal or not. A convenient algorithm was provided by Tarjan and Yannakakis (1984) and it is called *maximum cardinality search*; see also Cowell *et al.* (1999, section 4.4). If a graph is chordal this procedure provides additional useful information such as the cliques of the graphs as well as a perfect ordering of the cliques.

For two undirected graphs with common vertex set, $G = (V, E)$ and $G' = (V, E')$, we say that G is *larger than* G', and write $G' \subseteq G$, if $E' \subseteq E$; when the inclusion is strict, i.e., $E' \subset E$, we write $G' \subset G$. Frydenberg and Lauritzen (1989) showed that it is possible to explore the space of decomposable graphs by moving between neighboring decomposable models which differ exactly by one edge.

Proposition 4.6 *Let $G = (V, E)$ and $G' = (V, E')$ be two decomposable graphs such that $G' \subset G$ and $|E \backslash E'| = k$. Then there is sequence $G' = G_0 \subset \ldots \subset G_k = G$ of decomposable graphs that differ by exactly one edge.*

Frydenberg and Lauritzen (1989) also gave the following result concerning neighboring decomposable graphs that differ by one edge.

Proposition 4.7 *Let $G = (V, E)$ and $G' = (V, E')$ be two decomposable graphs such that $G' \subset G$ and G having exactly one more edge than G', i.e., $|E \backslash E'| = 1$. Then this edge is contained in exactly one clique C of G.*

4.8 Local Computation Properties

One key feature of graphical models is modularity; i.e., they exploit conditional independence relations to split complex systems into simpler modules. Consider the undirected graph model $M(G)$ for Y with independence graph $G = (V, E^\sim)$. If the pair of subsets $A, B \subseteq V$ is such that $A \cup B = V$ and $S = A \cap B$ separates $A \setminus B \neq \emptyset$ from $B \setminus A \neq \emptyset$ then it holds that $Y_{A \setminus B} \perp\!\!\!\perp Y_{B \setminus A} | Y_S$, and it follows from the definition of conditional independence (2.6) that the probability mass function of Y factorizes as

$$p(Y = i) = \frac{p(Y_A = i_A)p(Y_B = i_B)}{p(Y_S = i_S)} \quad \text{for all } i \in \mathcal{I}. \tag{4.16}$$

It is straightforward to see that this factorization can be equivalently stated in terms of expected cell counts $m(Y = i) = np(Y = i)$, and thus also the estimated expected cell counts factorize accordingly,

$$\hat{m}(Y = i) = \frac{\hat{m}(Y_A = i_A)\hat{m}(Y_B = i_B)}{\hat{m}(Y_S = i_S)} \quad \text{for all } i \in \mathcal{I}, \tag{4.17}$$

where $\{\hat{m}(Y_A = i_A)\}_{i_A \in \mathcal{I}_A}$ is obtained by marginalizing $\{\hat{m}(Y = i)\}_{i \in \mathcal{I}}$ over the variables $Y_{B \setminus A}$, and similarly for $\{\hat{m}(Y_B = i_B)\}_{i_B \in \mathcal{I}_B}$ and $\{\hat{m}(Y_S = i_S)\}_{i_S \in \mathcal{I}_S}$. Consider now the graph G_A, i.e., the subgraph of G induced by A. This graph can be used to define the undirected graph model $M(G_A)$ for Y_A, and we denote by $\{\hat{m}^A(Y_A = i_A)\}_{i_A \in \mathcal{I}_A}$ the ML estimates of the expected cell counts under $M(G_A)$. In general, the marginal table of estimates $\{\hat{m}(Y_A = i_A)\}_{i_A \in \mathcal{I}_A}$ is different from the estimate $\{\hat{m}^A(Y_A = i_A)\}_{i_A \in \mathcal{I}_A}$ computed under the model $M(G_A)$. One relevant exception is the case where (A, B) forms a decomposition of G because in this case the ML estimate in $M(G)$ can be obtained from the marginal vectors Y_A and Y_B by plugging in (4.17) the ML estimates computed under $M(G_A)$ and $M(G_B)$.

Theorem 4.8 *Let* $\{\hat{m}(Y = i)\}_{i \in \mathcal{I}}$ *be the ML estimates of the expected cell counts in the undirected graph model* $M(G)$. *If* (A, B) *forms a proper decomposition of* G *then it holds that*

$$\hat{m}(Y = i) = \frac{\hat{m}(Y_A = i_A)\hat{m}(Y_B = i_B)}{n(Y_S = i_S)} \quad \text{for all } i \in \mathcal{I},$$

and, furthermore, for every $i_A \in \mathcal{I}_A$ *and* $i_B \in \mathcal{I}_B$,

$$\hat{m}(Y_A = i_A) = \hat{m}^A(Y_A = i_A) \quad \text{and} \quad \hat{m}(Y_B = i_B) = \hat{m}^B(Y_B = i_B),$$

where $\{\hat{m}^A(Y_A = i_A)\}_{i_A \in \mathcal{I}_A}$ *and* $\{\hat{m}^B(Y_B = i_B)\}_{i_B \in \mathcal{I}_B}$ *are the ML estimates under* $M(G_A)$ *and* $M(G_B)$, *respectively.*

This theorem is a special case of a more general result proved in Lauritzen (1996, theorem 4.4). Theorem 4.8 states one of the most relevant features of graphical models, that is the *local computation* of several statistical quantities marginally on the component models $M(G_A)$ and $M(G_B)$. Consider the case where G^1 and G^0 are two graphs on the same vertex set such that $G^0 \subset G^1$. To see and example of this situation, let G^1 be the graph in Figure 4.1 and G^0 the graph obtained by removing from G^1 the edge $e \sim f$. Then the deviance difference (4.15) for comparing these two models is

$$\text{dev}(G^0|G^1) = 2\sum_{i \in \mathcal{I}} n(Y = i) \log \frac{\hat{m}_1(Y = i)}{\hat{m}_0(Y = i)}$$

where $\{\hat{m}_0(Y = i)\}_{i \in \mathcal{I}}$ and $\{\hat{m}_1(Y = i)\}_{i \in \mathcal{I}}$ are the ML estimates under $M(G^0)$ and $M(G^1)$, respectively. Assume now that (A, B) forms a proper decomposition of both G^0 and G^1, and that $G_A^0 = G_A^1$. More concretely, in the example above, the pair (A, B) that forms a decomposition of both graphs is $(\{a, b, c, d\}, \{c, d, e, f\})$. In this case, $G_{\{a,b,c,d\}}^0 = G_{\{a,b,c,d\}}^1$ is the graph in Figure 4.3A, whereas $G_{\{c,d,e,f\}}^1$ is the graph in Figure 4.3B and $G_{\{c,d,e,f\}}^0$ can be obtained from the latter by removing the edge $e \sim f$. It follows that we can apply Theorem 4.8 to obtain

$$\text{dev}(G^0|G^1) = 2\sum_{i \in \mathcal{I}} n(Y = i) \log \frac{\left(\dfrac{\hat{m}_1(Y_A = i_A)\hat{m}_1(Y_B = i_B)}{n(Y_S = i_S)} \right)}{\left(\dfrac{\hat{m}_0(Y_A = i_A)\hat{m}_0(Y_B = i_B)}{n(Y_S = i_S)} \right)}$$

$$= 2\sum_{i \in \mathcal{I}} n(Y = i) \log \frac{\hat{m}_1(Y_A = i_A)\hat{m}_1(Y_B = i_B)}{\hat{m}_0(Y_A = i_A)\hat{m}_0(Y_B = i_B)}.$$

However, because $G_A^0 = G_A^1$ then also $M(G_A^0) = M(G_A^1)$ and therefore $\hat{m}_0(Y_A = i_A) = \hat{m}_1(Y_A = i_A)$ for every $i_A = \mathcal{I}_A$. Hence,

$$\text{dev}(G^0|G^1) = 2\sum_{i_B \in \mathcal{I}_B} n(Y_B = i_B) \log \frac{\hat{m}_1(Y_B = i_B)}{\hat{m}_0(Y_B = i_B)},$$

and we can conclude that the deviance difference can be computed locally with respect to the marginal distribution of Y_B; i.e.,

$$\text{dev}(G^0|G^1) = \text{dev}(G_B^0|G_B^1). \tag{4.18}$$

In fact, in the example described above a further simplification is possible because one can exploit the decomposition $(\{c,d,f\}, \{c,e,f\})$ of both $G^0_{\{c,d,e,f\}}$ and $G^1_{\{c,d,e,f\}}$ to localize the computation of the deviance on the submodels defined by $G^0_{\{c,e,f\}}$ and $G^1_{\{c,e,f\}}$.

The local computation in (4.18) has the advantage that it simplifies the computation of the deviance and, furthermore, it involves marginal tables with larger expected cell counts with respect to the full table and this has typically a positive effect on the quality of the approximation of the distribution of the deviance to its asymptotic chi-squared distribution.

Local Computation of the Log-linear Terms

We now turn to the log-linear parameters of the model and to the local computation of these quantities. The following theorem gives the decomposition of the log-linear terms implied by conditional independence.

Theorem 4.9   *Let λ^V be the log-linear parameter of the distribution of Y_V with probability parameter $\pi > 0$. Consider a pair of subsets A and B such that $A \cup B = V$, with both $A \setminus B$ and $B \setminus A$ nonempty. Furthermore, let λ^A, λ^B and $\lambda^{A \cap B}$ be the log-linear parameters of the distributions of Y_A, Y_B and $Y_{A \cap B}$, respectively. If $Y_{A \setminus B} \perp\!\!\!\perp Y_{B \setminus A} | Y_{A \cap B}$, then in the binary case it holds that for all $D \subseteq V$*

$$\lambda_D = \lambda^A_D + \lambda^B_D - \lambda^{A \cap B}_D \qquad (4.19)$$

where we use the convention that $\lambda^A_D = 0$ whenever $D \not\subseteq A$, and similarly for λ^B and $\lambda^{A \cap B}$. In the general, non-binary, case (4.19) has the form $\lambda_{jD} = \lambda^A_{jD} + \lambda^B_{jD} - \lambda^{A \cap B}_{jD}$ for all $D \subseteq V$ and $j \in \mathcal{J}$.

Proof.   We first consider the binary case. Notice that if D is such that both $D \cap A \neq \emptyset$ and $D \cap B \neq \emptyset$ then the result is trivially true because $\lambda_D = 0$ by Theorem 4.2, whereas $\lambda^A_D = \lambda^B_D = \lambda^{A \cap B}_D = 0$ by assumption. Hence, in the rest of this proof we can assume, without loss of generality, that $D \subseteq A$. The conditional independence relationship $Y_{A \setminus B} \perp\!\!\!\perp Y_{B \setminus A} | Y_{A \cap B}$ implies that $\pi_D = \pi^A_{D \cap A} \pi^B_{D \cap B} / \pi^S_{D \cap S}$, for every $D \subseteq V$, and therefore that

$$\lambda_D = \sum_{E \subseteq D} (-1)^{|D \setminus E|} \log \pi_E$$

$$= \sum_{E \subseteq D} (-1)^{|D \setminus E|} \left(\log \pi^A_E + \log \pi^B_{E \cap B} - \log \pi^S_{E \cap S} \right)$$

where $E \cap A = E$ because $E \subseteq D \subseteq A$. For the same reason $E \cap B = E \cap S$ so we can write

$$\lambda_D = \left\{ \sum_{E \subseteq D} (-1)^{|D \setminus E|} \log \pi_E^A \right\} + \sum_{E \subseteq D} (-1)^{|D \setminus E|} (\log \pi_{E \cap S}^B - \log \pi_{E \cap S}^S)$$

$$= \lambda_D^A + \sum_{E \subseteq D} (-1)^{|D \setminus E|} \left(\log \pi_{E \cap S}^B - \log \pi_{E \cap S}^S \right).$$

Hence,

$$\lambda_D = \lambda_D^A + \sum_{E_1 \subseteq D \setminus S} \sum_{E_2 \subseteq D \cap S} (-1)^{|D \setminus (E_1 \cup E_2)|} \left(\log \pi_{E_2}^B - \log \pi_{E_2}^S \right)$$

where we have used the fact that $(E_1 \cup E_2) \cap S = E_2$. We can now factorize $(-1)^{|D \setminus (E_1 \cup E_2)|}$ as $(-1)^{|(D \setminus S) \setminus E_1|} \times (-1)^{|(D \cap S) \setminus E_2|}$ to obtain

$$\lambda_D = \lambda_D^A + \left\{ \sum_{E_1 \subseteq D \setminus S} (-1)^{|(D \setminus S) \setminus E_1|} \right\}$$

$$\times \sum_{E_2 \subseteq D \cap S} (-1)^{|(D \cap S) \setminus E_2|} \left(\log \pi_{E_2}^B - \log \pi_{E_2}^S \right). \tag{4.20}$$

At this point there are two possible cases, i.e., either $D \setminus S \neq \emptyset$ or $D \setminus S = \emptyset$. If $D \setminus S \neq \emptyset$ then the first quantity within brackets is equal to zero because any finite, nonempty set has the same number of subsets of even as of odd cardinality. Hence, $\lambda_D = \lambda_D^A$ and the result follows because in this case both $\lambda_D^B = \lambda_D^S = 0$ by assumption because both $D \not\subseteq B$ and $D \not\subseteq S$. On the other hand, if $D \setminus S = \emptyset$ then the first quantity within brackets is equal to one and, furthermore, in this case $D \cap S = D$ so we can write (4.20) as

$$\lambda_D = \lambda_D^A + \sum_{E_2 \subseteq D} (-1)^{D \setminus E_2} \log \pi_{E_2}^B - \sum_{E_2 \subseteq D} (-1)^{D \setminus E_2} \log \pi_{E_2}^S$$

$$= \lambda_D^A + \lambda_D^B - \lambda_D^S.$$

We have thus shown the assertion of the theorem for the binary case. The extension to the non-binary case can be obtained by repeating the proof above for λ_j for every $j \in \mathcal{J}$. $\qquad \square$

Let G be an undirected graph with vertex set V so that $\mathcal{V}$ is the simplicial complex relative to G. If the subsets $A, B \subseteq V$ are such that

$A \cup B = V$, then the triple $(A \setminus B, S, B \setminus A)$, where $S = A \cap B$, forms a partition of V. The following lemma shows that if S separates $A \setminus B$ from $B \setminus A$ in G then there exists a partition of V that resembles that of V.

Lemma 4.10 *Consider the simplicial complex V relative to the graph $G = (V, E^\sim)$. For the subsets $A, B \subseteq V$ set $S = A \cap B$ and let A, B and S be the simplicial complexes relative to G_A, G_B and G_S, respectively. Then, if S separates $A \setminus B$ from $B \setminus A$ in G it holds that*

$$A \cup B = V \text{ and } A \cap B = S$$

so that the triple $(A \setminus B, S, B \setminus A)$ forms a partition of V.

Proof. We first notice that if a pair of subsets $Q, E \subseteq V$ is such that if $Q \subseteq E$ then it also holds that $Q \subseteq \mathcal{E}$, where Q and $\mathcal{E}$ are the simplicial complexes relative G_Q and G_E, respectively. By definition $A, B \subseteq V$ so that $A, B \subseteq V$ and therefore $A \cup B \subseteq V$. Hence, to show the equality $A \cup B = V$ it is sufficient to show the reverse inclusion, i.e., that if $D \in V$ then $D \in A \cup B$. Any set $D \in V$ is complete in G, and the fact that $A \cap B = S$ separates $A \setminus B$ from $B \setminus A$ in G implies that at least one of the two inclusions $D \subseteq A$ and $D \subseteq B$ is satisfied. This implies, in turn, that at least one of the two inclusions $D \in A$ and $D \in B$ is satisfied and therefore that $D \in A \cup B$, as required. We now show that $A \cap B = S$. Because $S \subseteq A$ and $S \subseteq B$ then $S \subseteq A$ and $S \subseteq B$ and therefore $S \subseteq A \cap B$. We now show the reverse implication, that is that if $D \in A \cap B$ then $D \in S$. Any set $D \in A \cap B$ is such that both $D \in A$ and $D \in B$ which, in turn, imply both $D \subseteq A$ and $D \subseteq B$. Hence, $D \subseteq A \cap B = S$, and as D is complete in G it is also complete in G_S and therefore $D \in S$, and this completes the proof. $\square$

The partition of V in Lemma 4.10 provides a way to write the result of Theorem 4.9 in vector form. If λ is the log-linear parameter of a random vector under $M(G)$ then we can write $\lambda = (\lambda_V, \lambda_{\bar{V}})$ where $\lambda_{\bar{V}} = 0$. Furthermore, by Lemma 4.10, λ_V can be partitioned into $\lambda_V = (\lambda_{A \setminus B}, \lambda_S, \lambda_{B \setminus A})$ and the result of Theorem 4.9 can be written as

$$\lambda_V = \begin{pmatrix} \lambda_{A \setminus B} \\ \lambda_S \\ \lambda_{B \setminus A} \end{pmatrix} = \begin{pmatrix} \lambda_{A \setminus B}^A \\ \lambda_S^A + \lambda_S^B - \lambda_S^S \\ \lambda_{B \setminus A}^B \end{pmatrix}. \tag{4.21}$$

In graphical models for random variables with multivariate Gaussian distribution the role of λ is played by the inverse of the variance and covariance matrix, and this parameter can be locally computed in a way that resembles the local computation of λ in (4.21) as shown, for example, in Roverato and Whittaker (1998).

When (A, B) forms a proper decomposition of G then also the ML estimate $\hat{\lambda}$ can be computed locally.

Corollary 4.11 *Let $\hat{\lambda}$ be the ML estimate of the log-linear parameter of the distribution of the binary vector Y_V, with probability parameter $\pi > 0$, in the undirected graph model $M(G)$. If (A, B) forms a proper decomposition of G then it holds that*

$$
\hat{\lambda}_V = \begin{pmatrix} \hat{\lambda}_{A\backslash B} \\ \hat{\lambda}_S \\ \hat{\lambda}_{B\backslash A} \end{pmatrix} = \begin{pmatrix} \hat{\lambda}^A_{A\backslash B} \\ \hat{\lambda}^A_S + \hat{\lambda}^B_S - \hat{\lambda}^S_S \\ \hat{\lambda}^B_{B\backslash A} \end{pmatrix}, \tag{4.22}
$$

where $\hat{\lambda}^A$, $\hat{\lambda}^B$ and $\hat{\lambda}^S$ are the ML estimates of the log-linear parameters under $M(G_A)$, $M(G_B)$ and $M(G_S)$, respectively.

Proof. This follows immediately from Theorems 4.8 and 4.9, and from equation (4.21). □

The results given in equations (4.21) and (4.22) can be immediately extended to the case of categorical variables with arbitrary number of levels.

4.9 Models for Decomposable Graphs

Undirected graph models defined by decomposable graphs are called *decomposable*. Decomposable models are important for various reasons, and one is that many theoretical and computational aspects are more tractable with decomposable models than with general models. One reason for that is that the modules of a decomposable graph are all complete and this confers additional advantages. In this section, we focus on the decomposition of the ML estimate of the model and refer to Cowell *et al.* (1999) for a more comprehensive treatment of this topic.

A fundamental property of decomposable graphs is that they can be recursively decomposed into their cliques, which are complete subsets.

This recursive decomposition into the cliques of the graph implies that decomposable graphical models can themselves be recursively decomposed into saturated models for subsets of variables. As shown in Section 4.7, if G is a decomposable graph then its cliques can be numbered to form a perfect sequence $(C_1, \ldots, C_k)$. For every $r = 2, \ldots, k$ the pair of subsets $C_1 \cup \ldots \cup C_{r-1}$ and C_r forms a decomposition of the subgraph $G_{C_1 \cup \ldots \cup C_r}$, giving the sequence of separators $(S_2, \ldots, S_k)$. The repeated application of the factorization (4.16) to this sequence of decompositions produces a factorization of the probability mass function of Y into terms relative to the marginal distributions for cliques and separators,

$$p(Y = i) = \frac{\prod_{r=1}^{k} p(Y_{C_r} = i_{C_r})}{\prod_{r=2}^{k} p(Y_{S_r} = i_{S_r})}.$$

Accordingly, the repeated application of (4.17) gives the formula

$$\hat{m}(Y = i) = \frac{\prod_{r=1}^{k} \hat{m}(Y_{C_r} = i_{C_r})}{\prod_{r=2}^{k} \hat{m}(Y_{S_r} = i_{S_r})}.$$

Because the set of the cliques of G are the generating class of the model, then it follows from (4.14) that the ML estimates of the expected cell counts can be computed from the minimal sufficient statistics as follows (see also Lauritzen, 1996, Proposition 4.18).

Proposition 4.12 *Let $\{\hat{m}(Y = i)\}_{i \in \mathcal{I}}$ be the ML estimates of the expected cell counts in the undirected graph model $M(G)$. If G is decomposable then it holds that*

$$\hat{m}(Y = i) = \frac{\prod_{r=1}^{k} n(Y_{C_r} = i_{C_r})}{\prod_{r=2}^{k} n(Y_{S_r} = i_{S_r})}$$

where $\{C_1, \ldots, C_k\}$ is the set of cliques of G and $\{S_2, \ldots, S_k\}$ the set of separators.

Proposition 4.12 specifies the result of Theorem 4.8 to the decomposable case and shows that there exists an explicit formula for the ML estimate of a decomposable model. When applied to a decomposable model, the IPF will converge to the ML estimate after a finite number of steps, as this is the case for the application of the IPF to the infant survival data in Section 4.6.

Consider now the comparison of two nested decomposable models through the deviance statistic. As stated in Proposition 4.6, there exists a path of decomposable models from the larger to the smaller model, formed by removing one edge at the time. Hence, we can restrict the attention to the comparison of the decomposable models with graphs $G^1 = (V, E^1)$ and $G^0 = (V, E^0)$ where G^1 has exactly one more edge than G^1. To see an example of this framework, assume that G^1 is the graph in Figure 4.5 and that G^0 is obtained from the latter by removing the edge $c \sim e$. As stated in Proposition 4.7, the edge removed from G belongs to a unique clique C of G and to no separator. As a consequence, the comparison of the two models can be localized on the marginal distribution of Y_C so that

$$\mathrm{dev}(G^0|G^1) = \mathrm{dev}(G^0_C|G^1_C). \tag{4.23}$$

The local computation of the deviance statistic in (4.23) follows immediately from the factorization of the two probability mass functions with respect to the cliques and separators of G^1, but a formal proof can be found in Dawid and Lauritzen (1993, section 6). Consider the example with the graph G^1 in Figure 4.5. The removed edge $c \sim e$ belongs to the clique $\{c, d, e\}$ and, thus, the comparison of the two models can be localized on the marginal distribution of $Y_{\{c,d,e\}}$ to obtain

$$\mathrm{dev}(G^0|G^1) = \mathrm{dev}(G^0_{\{c,d,e\}}|G^1_{\{c,d,e\}})$$

$$= 2 \sum_{i_{\{c,d,e\}} \in \mathcal{I}_{\{c,d,e\}}} n(Y_{\{c,d,e\}} = i_{\{c,d,e\}}) \log \frac{\hat{m}_1(Y_{\{c,d,e\}} = i_{\{c,d,e\}})}{\hat{m}_0(Y_{\{c,d,e\}} = i_{\{c,d,e\}})}.$$

However, if one notices that the graph in Figure 4.5 has cliques $\{\{a, b\}, \{b, c\}, \{c, d, e\}, \{d, e, f\}, \{d, e, g\}\}$ and separators $\{\{b\}, \{c\}, \{d, e\}, \{d, e\}\}$, and that the graph G^0 has cliques $\{\{a, b\}, \{b, c\}, \{c, d\}, \{d, e, f\}, \{d, e, g\}\}$ and separators $\{\{b\}, \{c\}, \{d\}, \{d, e\}\}$, then, by exploiting the factorization of the ML estimate of Proposition 4.12, the above deviance can expressed in terms of marginal cell counts

$$\mathrm{dev}(G^0|G^1) = 2 \sum_{i_{\{c,d,e\}} \in \mathcal{I}_{\{c,d,e\}}} n(Y_{\{c,d,e\}} = i_{\{c,d,e\}})$$

$$\times \log \frac{n(Y_{\{c,d,e\}} = i_{\{c,d,e\}}) n(Y_{\{d\}} = i_{\{d\}})}{n(Y_{\{c,d\}} = i_{\{c,d\}}) n(Y_{\{d,e\}} = i_{\{d,e\}})}.$$

More generally, if we denote by $u \sim v$ the edge removed from G^1 to obtain G^0, and the removed edge belongs to the clique C of G^1, then the deviance statistic for comparing the corresponding decomposable models has form

$$\text{dev}(G^0|G^1) = 2 \sum_{i_C \in \mathcal{I}_C} n(Y_C = i_C)\log \frac{n(Y_C = i_C)n(Y_{C\setminus\{u,v\}} = i_{C\setminus\{u,v\}})}{n(Y_{C\setminus\{u\}} = i_{C\setminus\{u\}})n(Y_{C\setminus\{v\}} = i_{C\setminus\{v\}})}.$$

As well as the deviance statistic, other inferential quantities can be locally computed in a similar fashion on saturated marginal models. For instance, Dawid and Lauritzen (1993) gave the local computation of the Bayes factor for model comparison in Bayesian inference, whereas Roverato and Whittaker (1998) considered the local computation of the Laplace approximation to the Bayes factor.

We now turn to the log-linear parameterization of decomposable models. Similarly to Proposition 4.12, which reproduces the result of Theorem 4.8 in the special case of decomposable model, we give below a proposition that specifies the results of Theorem 4.9 and Corollary 4.11. For notational convenience, this is presented for the binary case; the generalization to the polytomous case is straightforward.

Proposition 4.13 *Let $\hat{\lambda}$ be the ML estimate, under $M(G)$, of the log-linear parameter of the distribution of the binary vector Y_V, with probability parameter $\pi > 0$. If G is decomposable with cliques $\{C_1, \ldots, C_k\}$ and separators $\{S_2, \ldots, S_k\}$ then*

$$\lambda_D = \sum_{r=1}^{k} \lambda_D^{C_r} - \sum_{r=2}^{k} \lambda_D^{S_r} \text{ and } \hat{\lambda}_D = \sum_{r=1}^{k} \hat{\lambda}_D^{C_r} - \sum_{r=2}^{k} \hat{\lambda}_D^{S_r}$$

where for every $r = 1, \ldots, k$, λ^{C_r} is the log-linear parameter of the distributions of Y_C, and $\hat{\lambda}^{C_r}$ its ML estimate in the saturated model, and similarly for the separators. Here, we use the convention that $\lambda_D^C = 0$ whenever $D \not\subseteq C$, and similarly for λ^S.

Proof. This follows from the recursive application of Theorem 4.9 and equation (4.22) with respect to a perfect sequence of cliques of G. □

The corresponding local computation of the inverse variance and covariance matrix in Gaussian graphical models can be found, for example, in Roverato and Whittaker (1998).

4.10   Log-linear Models and the Exponential Family

The exponential family is a convenient and widely used unified family of distributions whose members share many important properties. Hierarchical log-linear models belong to the exponential family (see Geiger and Meek, 1998; Koski and Noble, 2009; Lauritzen, 1996, p. 76) and in the next section we will exploit some general properties of this family of distributions in the asymptotic analysis of log-linear models. Much of the power of the exponential family formalism derives from the insights that are obtained from considering different parameterizations for a given family and, in this context, a central role is played by the connection between the canonical parameter and the mean parameter of the family. For notational convenience we restrict the attention to the binary case, and in this section we provide an overview of the alternative exponential family representations of the joint distribution of a vector of binary variables. Then, we derive an exponential family representation of log-linear models in which the canonical parameter is the collection of log-linear terms. In this way, we obtain a framework that naturally suits the subclass of graphical models, and thus the derivation of the results can be obtained by applying the properties of the regular exponential family together with the tools provided by the theory of Möbius inversion.

4.10.1   *Basic Facts on the Theory of the Exponential Family*

We first review the exponential family theory as required for this text, and refer to Barndorff-Nielsen (1978) for a full account on exponential families; see also Gutiérrez-Peña and Smith (1997) and Barndorff-Nielsen (2014).

Let Y be a random vector whose probability distribution belongs to a regular exponential family, then the density function of Y, with respect to an appropriate σ-finite measure, can be written in the form

$$f(y\,;\theta) = b(y)\exp\{\theta^T t(y) - M(\theta)\}$$

for some nonnegative measurable function $b(\cdot)$. The vectors $\theta \in \Theta$ and $T = t(Y)$ are the *canonical parameter* and the *canonical statistic* of the family, respectively, and $M(\theta)$ is known as the *cumulant function*. The *mean parameter* of the family is

$$\mu = \mu(\theta) = E_\theta(T) = \frac{\delta M(\theta)}{\delta \theta},$$

whereas the function

$$V(\mu) = \operatorname{var}_\theta(T) = \frac{\delta^2 M(\theta)}{\delta\theta\,\delta\theta^T} = \frac{\delta\mu(\theta)}{\delta\theta^T}$$

defined for $\mu \in \mu(\Theta)$ is called the *variance function* of the family; see Davison (2003, pp. 169 and 170) and Gutiérrez-Peña and Smith (1997, sections 2.1 and 2.2). The importance of the variance functions stems from the fact that, together with its domain $\mu(\Theta)$, it characterizes the family within the class of natural exponential families (Morris, 1982, 1983).

A general result for the regular exponential family (Barndorff-Nielsen, 1978, p. 150) is that the asymptotic variance of the ML estimate of the canonical parameter can be obtained as the inverse of the variance of the canonical statistic,

$$\operatorname{avar}(\hat{\theta}) = \operatorname{var}_\theta(T)^{-1} = V(\mu)^{-1} = \frac{\delta\theta(\mu)}{\delta\mu^T},$$

where $\theta(\cdot)$ denotes the inverse of the mapping $\mu(\cdot)$; see also Davison (2003, p. 171) and Gutiérrez-Peña and Smith (1997, section 2.2.1).

4.10.2 The Cross-classified Bernoulli Distribution

A variable that may take only the values zero and one is said to follow a Bernoulli distribution. When dealing with the binary case we have assumed that the state space of Y_V is given by $\{0,1\}^{|V|}$, and in the literature the distribution of this random vector is known as the *multivariate* or *cross-classified* Bernoulli. Here we introduce some useful quantities and recall a common way, based on the entries of π, to write the probability function of a cross-classified Bernoulli distribution. This will be elaborated in the forthcoming subsections to obtain the required exponential family representations.

For a set $D \subseteq V$ consider the following two Bernoulli random variables,

$$\widetilde{Y}_D = \prod_{v\in D} Y_v \quad \text{and} \quad \widetilde{Y}_D^c = \prod_{v\in D}(1 - Y_v), \tag{4.24}$$

with the convention that $\widetilde{Y}_\emptyset = \widetilde{Y}_\emptyset^c = 1$; see also Consonni and Leucari (2006). We remark that the quantity $\widetilde{Y}_D$ depends on Y only through the subvector Y_D, and similarly for $\widetilde{Y}_D^c$. Notice also that, for any pair of subsets $D, E \subseteq V$, it holds that both $\widetilde{Y}_{D \cup E} = \widetilde{Y}_D \times \widetilde{Y}_E$ and $\widetilde{Y}^c_{D \cup E} = \widetilde{Y}^c_D \times \widetilde{Y}^c_E$. We now introduce the further Bernoulli random variable

$$U^V_D = \widetilde{Y}_D \times \widetilde{Y}^c_{V \setminus D} \tag{4.25}$$

so that, in summary, we have introduced three vectors of binary variables, namely $\widetilde{Y} = (\widetilde{Y}_D)_{D \subseteq V}$, $\widetilde{Y}^c = (\widetilde{Y}^c_D)_{D \subseteq V}$ and $U = U^V = (U^V_D)_{D \subseteq V}$. It is clear from (4.25) that $U^V_D = 1$ if and only if both $\widetilde{Y}_D = 1$ and $\widetilde{Y}^c_{V \setminus D} = 1$, and therefore if and only if the observed value of Y belongs to the cell indexed by D. This implies that $p(U_D = 1) = \pi_D$, and therefore $E(U_D) = \pi_D$ or, more generally, that $E(U) = \pi$. Furthermore, the vector U can be used to write the density mass function of Y as

$$p(Y_D = 1, Y_{V \setminus D} = 0) = \prod_{E \subseteq V} \pi_E^{u_E} \text{ for every } D \subseteq V, \tag{4.26}$$

where the entry $\pi_\emptyset$ is determined by the affine constraint

$$\pi_\emptyset = 1 - \sum_{D \subseteq V; D \neq \emptyset} \pi_D. \tag{4.27}$$

4.10.3 Exponential Family Representations of the Saturated Model

When the parameter π is unrestricted, apart from the sum-to-one constraint (4.27), the cross-classified Bernoulli distribution belongs to the regular exponential family, and its density mass function is typically represented in two alternative ways. These are given below, and we refer to Brown (1986, p. 4), Lauritzen (1996, p. 76), Gutiérrez-Peña and Smith (1997, sections 2.1 and 2.2) and Davison (2003, p. 169) for a more detailed account on this issue.

The formulation of the probability distribution of Y in (4.26) leads to a first possible exponential family representation of the cross-classified Bernoulli,

$$p(Y_D = 1, Y_{V \setminus D} = 0) = \exp\left(\sum_{E \subseteq V} u_E \log \pi_E\right) \text{ for every } D \subseteq V, \tag{4.28}$$

where U and $\log \pi$ are the canonical statistic and parameter respectively. This can be found, for instance, in Brown (1986, p. 4) and Lauritzen

(1996, p. 76). However, this exponential family representation is not full because of the affine constraint (4.27); see Brown (1986, p. 4). A full regular exponential family representation is typically obtained by setting $\phi_D = \log \pi_D - \log \pi_\varnothing$, for all $D \subseteq V$, and writing

$$p(Y_D = 1, Y_{V \setminus D} = 0) = \exp\left\{ \sum_{E \subseteq V} \phi_E u_E - M(\phi) \right\} \tag{4.29}$$

where

$$M(\phi) = -\log \pi_\varnothing = \log\left\{ 1 + \sum_{D \subseteq V; D \neq \varnothing} \exp(\phi_D) \right\}$$

see, e.g., Brown (1986, p. 5) and Davison (2003, p. 175). Notice that in (4.29) the canonical parameter is $\phi_\vec{a}$ whereas $U_\vec{a}$ is the canonical statistic because $\phi_\varnothing = 0$ and, accordingly, the mean parameter is $\pi_\vec{a} = E(U_\vec{a})$. Recall that we are using the notation introduced in Section 3.1.1 so that, for instance, $\pi_\vec{a}$ is the subvector obtained from π by removing the entry indexed by the empty set. For the computation of the variance function $V(\pi_\vec{a})$, and of the asymptotic variance of the ML estimate of the canonical parameter $\mathrm{avar}(\hat{\phi}_\vec{a}) = V(\pi_\vec{a})^{-1}$, it is convenient to introduce the matrix $\mathbb{P}$ that can be regarded as a bivariate representation of the probability parameter π. Formally, for a nonempty subset $A \subseteq V$ we set $\mathbb{P}^A = \mathrm{diag}(\pi^A)$ so that $\mathbb{P}^A$ is the matrix with zero off-diagonal entries and diagonal entries given by the probabilities

$$\mathbb{P}^A_{D,D} = \pi^A_D = p(Y_D = 1, Y_{A \setminus D} = 0) \quad \text{for every } D \subseteq A,$$

and it is straightforward to see that $|\mathbb{P}^A| = \prod_{D \subseteq A} \pi^A_D$. According with the notation introduced in Section 3.1.1, $\mathbb{P}^A_\vec{a}$ is the submatrix of $\mathbb{P}^A$ obtained by removing the row and the column indexed by the empty set; namely, $\mathbb{P}^A_\vec{a} = \mathrm{diag}(\pi^A_\vec{a})$. Furthermore, we adopt here the usual convention that we may omit the superscript when it is equal to V, so that $\mathbb{P} = \mathbb{P}^V$.

The variance function for the exponential family representation (4.29) and its determinant are

$$V(\pi_\vec{a}) = \mathrm{var}(U_\vec{a}) = \mathbb{P}_\vec{a} - \pi_\vec{a}\pi_\vec{a}^T \quad \text{and} \quad |V(\pi_\vec{a})| = \prod_{D \subseteq V} \pi_D, \tag{4.30}$$

respectively; see Agresti (2013, section 16.1.4) and Lauritzen (1996, p. 77). The asymptotic variance of $\hat{\phi}_{\aleph}$ and its determinant are therefore

$$\operatorname{avar}(\hat{\phi}_{\aleph}) = V(\pi_{\aleph})^{-1} = \mathbb{P}_{\aleph}^{-1} + \frac{1}{\pi_{\varnothing}} 1_{\aleph} 1_{\aleph}^{T} \qquad (4.31)$$

and

$$|\operatorname{avar}(\hat{\phi}_{\aleph})| = |\mathbb{P}^{-1}| = \prod_{D \subseteq V} \frac{1}{\pi_D}, \qquad (4.31a)$$

where $1_{\aleph}$ is the vector indexed by the nonempty subsets of V with all entries equal to one. The computations of the determinant in (4.30), as well as of the inverse matrix in (4.31) and of its determinant, are based on standard results of matrix algebra, which can be found in Graybill (1983, theorems 8.3.3 and 8.4.3). Notice that $|\operatorname{avar}(\hat{\phi}_{\aleph})| = |\mathbb{P}^{-1}|$, although $\operatorname{avar}(\hat{\phi}_{\aleph}) \neq \mathbb{P}^{-1}$.

To facilitate the connection between the different parameterizations and statistics given in this section, Table 4.3 details the relevant quantities involved in equations (4.26), (4.28) and (4.29) for the case $V = \{a, b, c\}$.

4.10.4 Exponential Family Representation of Hierarchical Log-linear Models

The parametric restrictions implied by a log-linear model do not correspond to the vanishing of some elements of either $\log \pi$ or ϕ and, consequently, for an arbitrary hierarchical log-linear model the representations (4.28) and (4.29) are not full. It is thus of interest to give an exponential family representation of hierarchical log-linear models in which the canonical parameter is the collection of log-linear terms. This can be obtained by exploiting the fact that there exists a linear relationship between $\widetilde{Y}$ and U, which can be written in terms of a Möbius inversion. The interested reader may refer to Gutiérrez-Peña and Smith (1997) for a review on the role of linear transformations in regular exponential families.

Lemma 4.14 *Let* $\widetilde{Y} = (\widetilde{Y}_D)_{D \subseteq V}$ *and* $U = (U_D)_{D \subseteq V}$ *the vectors defined in (4.24) and (4.25), respectively. Then for all* $D \subseteq V$,

$$\widetilde{Y}_D = \sum_{E \subseteq V \backslash D} U_{D \cup E} \text{ and } U_D = \sum_{E \subseteq V \backslash D} (-1)^{|E|} \widetilde{Y}_{D \cup E}$$

Table 4.3 Quantities involved in equations (4.26), (4.28) and (4.29) for $V = \{a, b, c\}$.

D	$\tilde{y}_D$	$\tilde{y}_{V\backslash D}^c$	u_D	π_D	ϕ_D
$\emptyset$	1	$(1-y_a)(1-y_b)(1-y_c)$	$(1-y_a)(1-y_b)(1-y_c)$	$p(0,0,0)$	0
$\{a\}$	y_a	$(1-y_b)(1-y_c)$	$y_a(1-y_b)(1-y_c)$	$p(1,0,0)$	$\log \frac{p(1,0,0)}{p(0,0,0)}$
$\{b\}$	y_b	$(1-y_a)(1-y_c)$	$(1-y_a)y_b(1-y_c)$	$p(0,1,0)$	$\log \frac{p(0,1,0)}{p(0,0,0)}$
$\{c\}$	y_c	$(1-y_a)(1-y_b)$	$(1-y_a)(1-y_b)y_c$	$p(0,0,1)$	$\log \frac{p(0,0,1)}{p(0,0,0)}$
$\{a,b\}$	y_ay_b	$(1-y_c)$	$y_ay_b(1-y_c)$	$p(1,1,0)$	$\log \frac{p(1,1,0)}{p(0,0,0)}$
$\{a,c\}$	y_ay_c	$(1-y_b)$	$y_a(1-y_b)y_c$	$p(1,0,1)$	$\log \frac{p(1,0,1)}{p(0,0,0)}$
$\{b,c\}$	y_by_c	$(1-y_a)$	$(1-y_a)y_by_c$	$p(0,1,1)$	$\log \frac{p(0,1,1)}{p(0,0,0)}$
$\{a,b,c\}$	$y_ay_by_c$	1	$y_ay_by_c$	$p(1,1,1)$	$\log \frac{p(1,1,1)}{p(0,0,0)}$

that in matrix form can be written as

$$\widetilde{Y} = \mathbb{Z}U \text{ and } U = \mathbb{M}\widetilde{Y}.$$

Proof. We first show that $\widetilde{Y}_D^c = \sum_{E \subseteq D}(-1)^{|E|}\widetilde{Y}_E$ for all $D \subseteq V$. This part of the proof is by induction on the cardinality of D. For $|D| = 0$ the result is trivially true because $D = \emptyset$ and $\widetilde{Y}_\emptyset^c = \widetilde{Y}_\emptyset = 1$. We now show that if the result is true for a subset $D \subset V$ with $0 \le |D| < |V|$ then it is also true for $D \cup \{v\}$ where $v \in V \setminus D$. It follows immediately from the definition of $\widetilde{Y}^c$ that $\widetilde{Y}_{D \cup \{v\}}^c = \widetilde{Y}_D^c \times (1 - Y_v)$. Hence,

$$\widetilde{Y}_{D \cup \{v\}}^c = \widetilde{Y}_D^c \times (1 - Y_v)$$

$$= \widetilde{Y}_D^c - Y_v \times \widetilde{Y}_D^c$$

$$= \sum_{E \subseteq D}(-1)^{|E|}\widetilde{Y}_E - Y_v \times \sum_{E \subseteq D}(-1)^{|E|}\widetilde{Y}_E \tag{4.32}$$

$$= \sum_{E \subseteq D}(-1)^{|E|}\widetilde{Y}_E + \sum_{E \subseteq D}(-1)^{|E \cup \{v\}|}\widetilde{Y}_{E \cup \{v\}} \tag{4.33}$$

$$= \sum_{E \subseteq D \cup \{v\}; v \notin E}(-1)^{|E|}\widetilde{Y}_E + \sum_{E \subseteq D \cup \{v\}; v \in E}(-1)^{|E|}\widetilde{Y}_E$$

$$= \sum_{E \subseteq D \cup \{v\}}(-1)^{|E|}\widetilde{Y}_E \tag{4.34}$$

where in (4.32) we have applied the induction assumption, whereas in (4.33) we have used the facts that $(-1) \times (-1)^{|E|} = (-1)^{|E \cup \{v\}|}$ because $v \notin E$ and that $Y_v \times \widetilde{Y}_E = \widetilde{Y}_{E \cup \{v\}}$.

By definition, $U_D = \widetilde{Y}_D \widetilde{Y}_{V \setminus D}^c$ and by applying (4.34) we obtain

$$U_D = \widetilde{Y}_D \sum_{E \subseteq V \setminus D}(-1)^{|E|}\widetilde{Y}_E = \sum_{E \subseteq V \setminus D}(-1)^{|E|}\widetilde{Y}_{D \cup E}$$

which is one of the four stated equalities. The remaining three equalities follow immediately from Corollary 3.6. $\qquad \square$

The linear transformation in Lemma 4.14 can be used in (4.28) to transform the canonical parameter $\log \pi$ into λ. This is achieved by exploiting the fact that $\mathbb{M}\mathbb{Z}$ is the identity matrix so that

$$u^T \log \pi = u^T \mathbb{Z}^T \mathbb{M}^T \log \pi = (\mathbb{Z}u)^T(\mathbb{M}^T \log \pi) = \widetilde{y}^T \lambda.$$

Proposition 4.15 *If the distribution of the cross-classified Bernoulli random vector Y_V belongs to the hierarchical log-linear model $M(\mathcal{C})$, then*

$$p(Y_D = 1, Y_{V \setminus D} = 0) = \exp\left\{ \sum_{E \in \mathcal{C}_*} \lambda_E \widetilde{y}_E - M(\lambda) \right\} \tag{4.35}$$

where $M(\lambda) = -\lambda_\emptyset = -\log \pi_\emptyset$.

Proof. By recalling that $Y = \mathbb{Z}U$ by Lemma 4.14, that $\lambda = \mathrm{M}^T \log \pi$ and that $\mathbb{Z}^T \mathrm{M}^T$ is the identity matrix, it follows from (4.26) that

$$\log p(Y_D = 1, Y_{V \setminus D} = 0) = \sum_{E \subseteq V} u_E \log \pi_E$$

$$= u^T \log \pi$$

$$= u^T \mathbb{Z}^T \mathrm{M}^T \log \pi$$

$$= \widetilde{y}^T \lambda$$

$$= \sum_{E \subseteq V} \widetilde{y}_E^T \lambda_E. \tag{4.36}$$

As λ_E is equal to zero whenever $E \notin \mathcal{C}$, then (4.36) simplifies to

$$\log p(Y_D = 1, Y_{V \setminus D} = 0) = \sum_{E \in \mathcal{C}} \widetilde{y}_E \lambda_E = \sum_{E \in \mathcal{C}_*} \lambda_E \widetilde{y}_E + \lambda_\emptyset$$

because $\widetilde{y}_\emptyset = 1$, and this completes the proof. $\qquad \square$

The canonical statistic of the exponential family in (4.35) is $\widetilde{Y}_{\mathcal{C}_*}$ and we now turn to the mean parameter $E(\widetilde{Y}_{\mathcal{C}_*})$. Let $\mu = (\mu_D)_{D \subseteq V}$ be the vector with entries $\mu_D = p(Y_D = 1)$ for $D \subseteq V$ with $D \neq \emptyset$ and $\mu_D = 1$ for $D = \emptyset$. Clearly, $E(\widetilde{Y}_D) = \mu_D$ so that $\mu_{\mathcal{C}_*}$ is the mean parameter for the exponential family representation in (4.35). There is a close connection between π and μ. Indeed, it is easy to see that $\mu_D = \pi_D^D$ for any $D \neq \emptyset$, but, more interestingly, π and μ are connected through a Möbius inversion formula.

Proposition 4.16 *Let $\pi = (\pi_D)_{D \subseteq V}$ be the probability parameter of a binary random vector Y_V and let $\mu = (\mu_D)_{D \subseteq V}$ be such that $\mu_D = p(Y_D = 1)$ for $D \subseteq V$ with $D \neq \emptyset$ and $\mu_D = 1$ for $D = \emptyset$. Then, for all $D \subseteq V$*

$$\mu_D = \sum_{E \subseteq V \setminus D} \pi_{D \cup E} \quad and \quad \pi_D = \sum_{E \subseteq V \setminus D} (-1)^{|E|} \mu_{D \cup E},$$

that in matrix form can be written as

$$\mu = \mathbb{Z}\pi \quad and \quad \pi = \mathbb{M}\mu.$$

Proof. This proposition is an immediate consequence of Lemma 4.14. Each of the four equalities stated in the proposition can be obtained from the corresponding equalities in Lemma 4.14 by taking the expectation of both sides of the identity. $\square$

4.11 Modular Structure of the Asymptotic Variance of ML Estimates

Initially, local computation properties and modularity of graphical models were developed for probability distributions satisfying the Markov property with respect to an undirected graph. Subsequently, for the decomposable case, Dawid and Lauritzen (1993) introduced the *hyper Markov properties* for probability distributions defined at a *hyper* level, that is over the set of probability distributions forming a graphical model. They considered, for instance, the distribution of the ML estimates for the parameters of a graphical model as well as prior and posteriors distributions.

In this section, we consider the binary case and describe the asymptotic theory of hierarchical log-linear models in a formulation that allows one to fully exploit the specific features of graphical models. For hierarchical log-linear models this approach provides an alternative way to derive certain classical results of asymptotic theory given, for example, in Christensen (1997, section 12.3) and Agresti (2013, chapter 16). For the subclass of undirected graph models this approach allows us to describe the asymptotic theory of these models by giving explicit rules for the local computation of the asymptotic variance of ML estimates of the log-linear parameters.

Although our main focus is on the asymptotic distribution of the ML estimates, and more specifically on the asymptotic variance of $\hat{\lambda}$, it is worth recalling that the material presented here can be applied to obtain a modular representation of other relevant inferential quantities. In a Bayesian approach to inference, the Bayes factor (Jeffreys, 1961) is a fundamental tool for comparing competing hypotheses. However, the Bayes factor can be locally computed, in a similar fashion as the deviance difference in Section 4.5, only in the special case where the prior distribution fulfills the *strong hyper Markov property* (Dawid and Lauritzen, 1993). A widely used asymptotic version of the Bayes factor is its *Laplace approximation* (Kass and Raftery, 1995), and in this case local computations are possible also under a less stringent requirement for the prior distribution, known as the *weak* hyper Markov property (Dawid and Lauritzen, 1993; Roverato and Whittaker, 1998). On the other hand, the local computation of the Laplace approximation to the Bayes factor is based on a modular representation of the determinant of the observed information matrix. Because log-linear models are regular exponential families, the observed information matrix corresponds to the expected information matrix evaluated at $\hat{\lambda}$. Furthermore, it can be obtained as the inverse of the asymptotic variance of the ML estimate of λ for a single observation, and therefore the modular representation of the determinant of the information matrix can be obtained by direct application of the material described here. This modular representation of the determinant is also useful when the interest is in the noninformative Jeffreys prior (Jeffreys, 1946) because it leads to a factorization of this quantity in a way similar to that of hyper Markov laws.

In the rest of this section we assume, without loss of generality, that the sample size is equal to one, so that the above asymptotic variance coincides with the inverse of the expected Fisher information matrix for the canonical parameter.

4.11.1 The Variance Function and the Asymptotic Variance of ML Estimates

This section is devoted to the variance function $V(\mu_{C_a})$ and the asymptotic variance of $\hat{\lambda}_{C_a}$. Special attention is paid to the description of the interesting mathematical relationship existing between these two quantities.

We have considered three alternative parameterizations of the cross-classified Bernoulli distribution, namely, μ_a, λ_a and π_a. There is a one-to-one relationship between every pair of these parameters. However, in the exponential family theory a key role is played by the functional relationship between the mean parameter μ_a and the canonical parameter $\lambda_{\mathscr{E}}$ because this determines the structure both of the variance function $V(\mu_{\mathscr{E}})$ and of the asymptotic variance of $\hat{\lambda}_{\mathscr{E}}$, which is a quantity of fundamental importance in statistical inference. The theory developed in the previous sections allows us to split the functional relationship between $\mu_{\mathscr{E}}$ and $\lambda_{\mathscr{E}}$ into easily manageable building blocks. This is achieved by considering the vectors μ and λ and using equation (4.1) and Propositions 4.16 to obtain λ from μ by means of a sequence of remarkably simple bijective functions, mainly involving Möbius inversion operations. The probability parameter π turns out to be the intermediate step of this decomposition,

$$\mu \quad \overset{\text{Möb. inv.}}{\Longleftrightarrow} \quad \pi \quad \Longleftrightarrow \quad \log \pi \quad \overset{\text{Möb. inv.}}{\Longleftrightarrow} \quad \lambda. \qquad (4.37)$$

Each of the vectors involved in (4.37) is indexed by the subsets of V, i.e., by the elements of $\mathcal{P}(V)$. We now turn to matrices indexed by the elements of $\mathcal{P}(V) \times \mathcal{P}(V)$. More specifically, we consider the matrix $\mathbb{P}$ introduced in Section 4.10.3 and the matrix $\mathbb{V}$ with entries

$$\mathbb{V}_{D,D'} = \mu_{D \cup D'} \text{ for all } D, D' \subseteq V.$$

Notice that the main diagonal of $\mathbb{P}$ is π and the main diagonal of $\mathbb{V}$ is μ. We show below that the inverse $\mathbb{V}^{-1}$ can be computed by means of a sequence of simple bijective functions that closely resembles (4.37), namely,

$$\mathbb{V} \quad \overset{\text{Möb. inv.}}{\Longleftrightarrow} \quad \mathbb{P} \quad \Longleftrightarrow \quad \mathbb{P}^{-1} \quad \overset{\text{Möb. inv.}}{\Longleftrightarrow} \quad \mathbb{V}^{-1}, \qquad (4.38)$$

where the Möbius inversion formula used here is that of Corollary 3.7.

There is an interesting parallelism between (4.37) and (4.38): a dimensionality reduction of μ leads to the mean parameter $\mu_{\mathscr{E}}$, and Proposition 4.19 shows that a dimensionality reduction of $\mathbb{V}$ leads to the variance function $V(\mu_{\mathscr{E}})$. On the other hand, a dimensionality reduction of λ leads to the canonical parameter $\lambda_{\mathscr{E}}$, and Theorem 4.21 shows that a dimensionality reduction of $\mathbb{V}^{-1}$ leads to avar$(\hat{\lambda}_{\mathscr{E}})$, the asymptotic variance of $\hat{\lambda}_{\mathscr{E}}$. The decomposition of the functional relationship between $\mathbb{V}$ and its inverse given in (4.38) is of theoretical interest as well as of practical usefulness because, by linking both $\mathbb{V}$ and $\mathbb{V}^{-1}$ with the diagonal matrix $\mathbb{P}$, it allows to derive an explicit formulations of the entries and of the determinant of both $V(\mu_{\mathscr{E}})$ and avar$(\hat{\lambda}_{\mathscr{E}})$.

As shown in Proposition 4.16, μ can be obtained from π by applying the Möbius inversion (a') of Corollary 3.6 with respect to $\mathcal{P}(V)$. As well as $\mathbb{P}$ can be regarded as a two-dimensional representation of π, the matrix $\mathbb{V}$ can be regarded as a two-dimensional representation of the vector μ because $\mathbb{V}$ can be obtained from $\mathbb{P}$ by applying the same Möbius inversion operation with respect to $\mathcal{P}(V) \times \mathcal{P}(V)$.

Proposition 4.17 *The functional relationship between the matrices $\mathbb{V}$ and $\mathbb{P}$ is as follows*

$$\mathbb{V} = \mathbb{Z}\mathbb{P}\mathbb{Z}^T \quad and \quad \mathbb{P} = \mathbb{M}\mathbb{V}\mathbb{M}^T.$$

Proof. By point (a'') of Corollary 3.7 it follows that the equalities $\mathbb{V} = \mathbb{Z}\mathbb{P}\mathbb{Z}^T$ *is satisfied if and only if*

$$\mathbb{V}_{D,D'} = \sum_{(E,E') \geq (D,D')} \mathbb{M}_{E,E'} \quad \text{for every } D, D' \subseteq V \tag{4.39}$$

and we now prove that (4.39) holds. By definition, $\mathbb{P}'_{E,E'} = \pi_E$ whenever $E = E'$, and it is equal to zero otherwise; hence, for every $D, D' \subseteq V$ it holds that

$$\sum_{(E,E') \geq (D,D')} \mathbb{P}_{E,E'} = \sum_{(E,E) \geq (D,D')} \pi_E.$$

Furthermore, by noticing that $(E, E) \geq (D, D')$ if and only if $E \geq D \cup D'$, and therefore if and only if $D \cup D' \subseteq E$, for every $D, D' \subseteq V$ we can write

$$\sum_{(E,E) \geq (D,D')} \pi_E = \sum_{F \subseteq V \setminus (D \cup D')} \pi_{D \cup D' \cup F} = \mu_{D \cup D'}$$

where the last identity follows from Proposition 4.16. This proves (4.39) because $\mathbb{V}_{D,D'} = \mu_{D \cup D'}$ by definition.

The equality $\mathbb{P} = \mathbb{M}\mathbb{V}\mathbb{M}^T$ follows from point (A'') of Corollary 3.7. $\qquad\square$

We have thus proved the first Möbius inversion in (4.38) that, interestingly, coincides with the Cholesky decomposition of $\mathbb{V}$. The second Möbius inversion is an immediate consequence of Proposition 4.17 and coincides with the Cholesky decomposition of $\mathbb{V}^{-1}$.

Corollary 4.18 *The functional relationship between the matrices $\mathbb{V}^{-1}$ and $\mathbb{P}^{-1}$ is as follows*

$$\mathbb{V}^{-1} = \mathbb{M}^T \mathbb{P}^{-1} \mathbb{M} \quad and \quad \mathbb{P}^{-1} = \mathbb{Z}^T \mathbb{V}^{-1} \mathbb{Z}.$$

Proof. This is an immediate consequence of Proposition 4.17 because $\mathbb{V} = \mathbb{Z}\mathbb{P}\mathbb{Z}^T$ implies $\mathbb{V}^{-1} = (\mathbb{Z}\mathbb{P}\mathbb{Z}^T)^{-1} = \mathbb{M}^T\mathbb{P}^{-1}\mathbb{M}$, and similarly for the second identity.   □

We have thus shown the validity of (4.38) and in the following we describe the connection between $\mathbb{V}$ and the variance function $V(\mu_{\&})$ and between $\mathbb{V}^{-1}$ and the asymptotic variance of $\hat{\lambda}_{\&}$.

4.11.2 Variances in the Saturated Model

The computation of the variance function and of the asymptotic variance of the ML estimate requires the notion of partial matrix. A matrix $\mathbb{G}$ indexed by the subsets of V can be partitioned with respect to any nonempty subset $C \subseteq \mathcal{P}(V)$,

$$\mathbb{G} = \begin{bmatrix} \mathbb{G}_{CC} & \mathbb{G}_{C\bar{C}} \\ \mathbb{G}_{\bar{C}C} & \mathbb{G}_{\bar{C}\bar{C}} \end{bmatrix}$$

where $\bar{C} = \mathcal{P}(V)\setminus C$. If $\mathbb{G}$ has full rank then the partial matrix of $\mathbb{G}$ with respect to C is defined as

$$\mathbb{G}_{CC|\bar{C}} = \mathbb{G}_{CC} - \mathbb{G}_{C\bar{C}}\mathbb{G}_{\bar{C}\bar{C}}^{-1}\mathbb{G}_{\bar{C}C}.$$

The saturated model only involves *partial matrices* where $C = \mathcal{P}(V)\setminus\emptyset$ and we use the shorthand $\mathbb{G}_{|e}$ to denote them.

The Variance Function

We can now give the variance function for the saturated model.

Proposition 4.19 *Let Y_V be a cross-classified Bernoulli random vector with probability parameter $\pi > 0$. If the probability distribution of Y_V belongs to the saturated model, then the variance function in the exponential family form (4.35) is*

$$V(\mu_{\&}) = \mathbb{V}_{|e}$$

and its entries are

$$V(\mu_{\&})_{D,D'} = \mu_{D\cup D'} - \mu_D\mu_{D'}$$

for all $D, D' \subseteq V$ with $D, D' \neq \emptyset$. Furthermore, the determinant of $V(\mu_{\&})$ is given by

$$|V(\mu_{\&})| = |\mathbb{V}| = |\mathbb{P}| = \prod_{D \subseteq V} \pi_D.$$

Proof. We first compute the entries of the variance function. By definition, $V(\mu_{\&}) = \text{var}(Y_{\&})$ so that the entry of $V(\mu_{\&})$ indexed by $D, D' \subseteq V$ with $D, D' \neq \varnothing$ can be computed as follows,

$$
\begin{aligned}
V(\mu_{\&})_{D,D'} &= \text{cov}(\tilde{Y}_D, \tilde{Y}_{D'}) \\
&= E(\tilde{Y}_D \tilde{Y}_{D'}) - E(\tilde{Y}_D)E(\tilde{Y}_{D'}) \\
&= E(\tilde{Y}_{D \cup D'}) - E(\tilde{Y}_D)E(\tilde{Y}_{D'}) \\
&= \mu_{D \cup D'} - \mu_D \mu_{D'}.
\end{aligned}
\tag{4.40}
$$

We now show that $V(\mu_{\&}) = \mathbb{V}_{|e}$. The first row and column of $\mathbb{V}$ are indexed by the empty set and, therefore, the matrix $\mathbb{V}$ can be written in a partitioned form as

$$
\mathbb{V} = \begin{bmatrix} \mathbb{V}_{\varnothing,\varnothing} & \mu_{\&}^T \\ \mu_{\&} & \mathbb{V}_{\&} \end{bmatrix}
$$

and, as $\mathbb{V}_{\varnothing,\varnothing} = \mu_{\varnothing} = 1$, it holds that $\mathbb{V}_{|e} = \mathbb{V}_{\&} - \mu_{\&}\mu_{\&}^T$. From the latter, it is easy to check that the entry of $\mathbb{V}_{|e}$ indexed by $D, D' \subseteq V$ with $D, D' \neq \varnothing$ is equal to (4.40) and this implies that $V(\mu_{\&}) = \mathbb{V}_{|e}$. Finally, we turn to the determinant of $V(\mu_{\&})$. The determinant of $\mathbb{V}$ can be factorized as $|\mathbb{V}| = |\mathbb{V}_{|e}||\mathbb{V}_{\varnothing,\varnothing}$ (see Lütkepol, 1996, p. 147, eqn. 4) so that $|\mathbb{V}| = |\mathbb{V}_{|e}|$ because $\mathbb{V}_{\varnothing,\varnothing} = 1$. Hence, by recalling that $|\mathbb{Z}| = 1$, we obtain

$$|V(\mu_{\&})| = |\mathbb{V}_{|e}| = |\mathbb{V}| = |\mathbb{Z}\mathbb{P}\mathbb{Z}^T| = |\mathbb{P}| = \prod_{D \subseteq V} \pi_D.$$

$\square$

As a consequence of Proposition 4.19 it is also possible to give an alternative, very compact, proof of the determinants in (4.30) and (4.31a).

Corollary 4.20 *The variance function of the exponential family in (4.29) has determinant* $|V(\pi_{\&})| = |V(\mu_{\&})| = \prod_{D \subseteq V} \pi_D.$

Proof. By definition $V(\pi_{\&}) = \text{var}(U_{\&})$ and because $U_{\&} = \mathbb{M}_{\&}Y_{\&}$ then

$$
\begin{aligned}
|V(\pi_{\&})| &= |\text{var}(\mathbb{M}_{\&}Y_{\&})| = |\mathbb{M}_{\&}\text{var}(Y_{\&})\mathbb{M}_{\&}^T| = |\mathbb{M}_{\&}V(\mu_{\&})\mathbb{M}_{\&}^T| \\
&= |V(\mu_{\&})| = \prod_{D \subseteq V} \pi_D.
\end{aligned}
$$

$\square$

The Asymptotic Variance of the ML Estimate

The asymptotic variance of $\hat{\lambda}_{\&}$ can be computed as the inverse of the variance function, $\mathrm{avar}(\hat{\lambda}_{\&}) = (\mathbb{V}_{|e})^{-1}$. Consequently, by using the results for the inverse of a partitioned matrix it follows immediately that $\mathrm{avar}(\hat{\lambda}_{\&})$ is a submatrix of $\mathbb{V}^{-1}$. We show in Theorem 4.21 that this has two important implications for the asymptotic variance. Firstly, it relates this quantity to a triangular decomposition involving the matrix $\mathbb{M}$, which is upper triangular with unit diagonal, and the diagonal matrix $\mathbb{P}^{-1}$. Secondly, it allows one to exploit the Möbius inversion in Corollary 4.18 to obtain an explicit formulation of the entries of $\mathrm{avar}(\hat{\lambda}_{\&})$.

Theorem 4.21 *Let Y_V be a cross-classified Bernoulli random vector with probability parameter $\pi > 0$. Then the asymptotic variance of the ML estimate of $\lambda_{\&}$ in the saturated model is*

$$\mathrm{avar}(\hat{\lambda}_{\&}) = (\mathbb{M}^T \mathbb{P}^{-1} \mathbb{M})_{\&} \tag{4.41}$$

and has entries

$$\mathrm{avar}(\hat{\lambda}_{\&})_{D,D'} = \sum_{E \subseteq D \cap D'} (-1)^{|D \setminus E| + |D' \setminus E|} \frac{1}{\pi_E} \tag{4.42}$$

for all $D, D' \subseteq V$ with $D, D' \neq \emptyset$. Furthermore, its determinant is

$$|\mathrm{avar}(\hat{\lambda}_{\&})| = |\mathbb{P}^{-1}| = \prod_{D \subseteq V} \frac{1}{\pi_D}. \tag{4.43}$$

Proof. Equation (4.41) can be derived by recalling that $\mathrm{avar}(\hat{\lambda}_{\&}) = V(\mu_{\&})^{-1}$ and then applying, in turn, Proposition 4.19, the rules for the inverse of a partitioned matrix (see Lütkepol, 1996, p. 29, eqn. 1) and, finally, Corollary 4.18 as follows

$$\mathrm{avar}(\hat{\lambda}_{\&}) = V(\mu_{\&})^{-1} = (\mathbb{V}_{|e})^{-1} = (\mathbb{V}^{-1})_{\&} = (\mathbb{M}^T \mathbb{P}^{-1} \mathbb{M})_{\&}.$$

In order to show (4.42) it is sufficient to compute the entries of $\mathbb{V}^{-1}$ because (4.41) implies that $\mathrm{avar}(\hat{\lambda}_{\&})_{D,D'} = (\mathbb{V}^{-1})_{D,D'}$ for every $D, D' \subseteq V$ with $V \neq \emptyset$. As $(\mathbb{V}^{-1})_{\&} = \mathbb{M}^T \mathbb{P}^{-1} \mathbb{M}$ by Corollary 4.18, we can apply Corollary 3.7 to write the entry of $\mathbb{V}^{-1}$ indexed by $D, D' \subseteq V$ as

$$(\mathbb{V}^{-1})_{D,D'} = \sum_{(E,E')\in\mathcal{P}(V)\times\mathcal{P}(V)} m_{\mathcal{P}(V)\times\mathcal{P}(V)}\{(E,E'),(D,D')\}(\mathbb{P}^{-1})_{E,E'}$$

$$= \sum_{E\in\mathcal{P}(V)} m_{\mathcal{P}(V)\times\mathcal{P}(V)}\{(E,E),(D,D')\}\frac{1}{\pi_E} \tag{4.44}$$

$$= \sum_{E\subseteq V} (-1)^{|D\setminus E|+|D'\setminus E|}1(E\subseteq D')1(E\subseteq D)\frac{1}{\pi_E} \tag{4.45}$$

$$= \sum_{E\subseteq D\cap D'} (-1)^{|D\setminus E|+|D'\setminus E|}\frac{1}{\pi_E} \tag{4.46}$$

where (4.44) follows from the fact that $(\mathbb{P}^{-1})_{E,E'} = \pi_E^{-1}$ whenever $E = E'$ and $(\mathbb{P}^{-1})_{E,E'} = 0$ otherwise. Furthermore, (4.45) follows from (3.14), whereas (4.46) follows from the fact that $1(E\subseteq D')1(E\subseteq D) = 1$ if $E \subseteq D\cap D'$ and it is equal to zero otherwise.

Finally, (4.43) is an immediate consequence of Proposition 4.19 because $|\mathrm{avar}(\hat{\lambda}_{\text{æ}})| = |V(\mu_{\text{æ}})^{-1}| = |V(\mu_{\text{æ}})|^{-1}$. $\qquad\square$

4.11.3 Variances in Hierarchical Log-linear Models

We now turn to an arbitrary log-linear model with expanded generating class $\mathcal{C}$.

The Variance Function
We first consider the variance function.

Proposition 4.22 *Let Y_V be a cross-classified Bernoulli random vector with probability parameter $\pi > 0$. If the probability distribution of Y_V belongs to the log-linear model $M(\mathcal{C})$, then the variance function in the exponential family form (4.35) is*

$$V(\mu_{C_{\text{æ}}}) = [\mathbb{V}c]_{|e}.$$

and, furthermore, $|V(\mu_{C_{\text{æ}}})| = |\mathbb{V}c|$.

Proof. The first row and column of $\mathbb{V}_C$ are indexed by the empty set and, therefore, the matrix $\mathbb{V}$ can be written in a partitioned form as

$$\mathbb{V}_C = \begin{bmatrix} \mathbb{V}_{\emptyset,\emptyset} & \mu_{C_{\text{æ}}}^T \\ \mu_{C_{\text{æ}}} & \mathbb{V}_{C_{\text{æ}}} \end{bmatrix}$$

so that, by recalling that $\mathbb{V}_{\varnothing,\varnothing} = \mu_\varnothing = 1$ we obtain $[\mathbb{V}_c]_{|e} = \mathbb{V}_{C_\mathtt{a}} - \mu_{C_\mathtt{a}}\mu_{C_\mathtt{a}}^T$. The result follows because $V(\mu_{C_\mathtt{a}}) = \mathrm{var}(Y_{C_\mathtt{a}}) = [\mathrm{var}(Y_\mathtt{a})]_{C_\mathtt{a}} = [\mathbb{V}_{|e}]_{C_\mathtt{a}} = [\mathbb{V}_\mathtt{a} - \mu_\mathtt{a}\mu_\mathtt{a}^T]_{C_\mathtt{a}} = \mathbb{V}_{C_\mathtt{a}} - \mu_{C_\mathtt{a}}\mu_{C_\mathtt{a}}^T$. The determinant of $\mathbb{V}_C$ can be factorized as $|\mathbb{V}_C| = |[\mathbb{V}_c]_{|e}|\mathbb{V}_{\varnothing,\varnothing}$ (see Lütkepol, 1996, p. 147, eqn. 4) so that $|\mathbb{V}_C| = |[\mathbb{V}_c]_{|e}| = |V(\mu_{C_\mathtt{a}})|$ because $\mathbb{V}_{\varnothing,\varnothing} = 1$. □

The Asymptotic Variance of the ML Estimate

The asymptotic variance of $\hat{\lambda}_{C_\mathtt{a}}$ can be easily derived from Proposition 4.22 by applying the rule for the inversion of a partitioned matrix.

Proposition 4.23 *Let Y_V be a cross-classified Bernoulli random vector with probability parameter $\pi > 0$. Then the asymptotic variance of the ML estimate of $\lambda_{C_\mathtt{a}}$ under the log-linear model $M(C)$ is*

$$\mathrm{avar}(\hat{\lambda}_{C_\mathtt{a}}) = \left[(M^T \mathbb{P}^{-1} M)_{CC|\bar{C}}\right]_\mathtt{a}$$

and, furthermore, $|\mathrm{avar}(\hat{\lambda}_{C_\mathtt{a}})| = |(M^T \mathbb{P}^{-1} M)_{CC|\bar{C}}|.$

Proof. We first note that, by applying Proposition 4.17, Corollary 4.18 and the rules for the inverse of a partitioned matrix (see Lütkepol, 1996, p. 29, eqn. 1), we can write

$$\mathbb{V}_{CC} = \left[\mathbb{Z}\mathbb{P}\mathbb{Z}^T\right]_{CC} = \left[(M^T \mathbb{P}^{-1} M)^{-1}\right]_{CC} = \left[(M^T \mathbb{P}^{-1} M)_{CC|\bar{C}}\right]^{-1}.$$

$$(4.47)$$

Hence, the stated result can be derived by recalling that $\mathrm{avar}(\hat{\lambda}_{C_\mathtt{a}}) = V(\mu_{C_\mathtt{a}})^{-1}$ and then applying, in turn, Proposition 4.22, the rules for the inverse of a partitioned matrix and, finally, (4.47)

$$\mathrm{avar}(\hat{\lambda}_{C_\mathtt{a}}) = V(\mu_{C_\mathtt{a}})^{-1} = ([\mathbb{V}_c]_{|e})^{-1} = \left[(\mathbb{V}_C)^{-1}\right]_\mathtt{a} = \left[(M^T \mathbb{P}^{-1} M)_{CC|\bar{C}}\right]_\mathtt{a}.$$

Furthermore, as a consequence of Proposition 4.22 and of (4.47) we obtain $|\mathrm{avar}(\hat{\lambda}_{C_\mathtt{a}})| = |V(\mu_{C_\mathtt{a}})^{-1}| = |(\mathbb{V}_C)^{-1}| = |(M^T \mathbb{P}^{-1} M)_{CC|\bar{C}}|.$ □

We remark that alternative formulae for $\mathrm{avar}(\hat{\lambda}_{C_\mathtt{a}})$ are available in the literature. For instance, $\mathrm{avar}(\hat{\lambda}_{C_\mathtt{a}})$ can be obtained from $V(\mu_{C_\mathtt{a}})$ as follows

$$V(\mu_{C_\mathtt{a}}) = \mathrm{var}(\widetilde{Y}_{C_\mathtt{a}}) = \mathrm{var}(\mathbb{Z}_{C_\mathtt{a},\mathcal{P}(V)}U) = \mathbb{Z}_{C_\mathtt{a},\mathcal{P}(V)}(\mathbb{P} - \pi\pi^T)\mathbb{Z}_{C_\mathtt{a},\mathcal{P}(V)}^T$$

so that

$$\text{avar}(\hat{\lambda}_{C_{\circledast}}) = V(\mu_{C_{\circledast}})^{-1} = \left\{ \mathbb{Z}_{C_{\circledast}, \mathcal{P}(V)} (\mathbb{P} - \pi\pi^T) \mathbb{Z}_{C_{\circledast}, \mathcal{P}(V)}^T \right\}^{-1}; \qquad (4.48)$$

see also Agresti (2013, eqn. 16.22). In Proposition 4.23 we give an alternative formulation of $\text{avar}(\hat{\lambda}_{\circledast})$ thereby providing, in explicit form, the matrix inversion required in (4.48). Such formulation, based on a Cholesky decomposition involving a Möbius matrix and a diagonal matrix of probabilities, is mathematically appealing, and we deem that it will be useful in the development of other aspects of these models. Indeed, we have exploited such Cholesky decomposition in Theorem 4.21 to give, in the saturated case, an explicit formulation of the entries of $\text{avar}(\hat{\lambda}_{\circledast})$ as well as of its determinant. The next section provides generalisation of Theorem 4.21 to the case in which the log-linear model considered is a decomposable graphical model.

4.11.4 Decompositions and Decomposable Models

We consider now the case where the distribution of Y_V belongs to the graphical model $M(G)$. If (A, B) is a decomposition of G, then by the collapsibility of graphical models (Asmussen and Edwards, 1983) it holds that the distribution of Y_A belongs to $M(G_A)$ and, similarly, the distribution of Y_B belongs to $M(G_B)$. The corresponding log-linear parameters are λ_A^A and λ_B^B, respectively, and Proposition 4.23 allows us to write the asymptotic variances of the ML estimates under these two models as

$$\text{avar}(\hat{\lambda}_{A_{\circledast}}^A) = \left[\left(M_A^T \mathbb{P}^{-A} M_A \right)_{A A | \overline{A}} \right]_{\circledast} \qquad (4.49)$$

and

$$\text{avar}(\hat{\lambda}_{B_{\circledast}}^B) = \left[\left(M_B^T \mathbb{P}^{-B} M_B \right)_{B B | \overline{B}} \right]_{\circledast} \qquad (4.49a)$$

where, in this case, $\overline{A} = \mathcal{P}(A) \backslash \mathcal{A}$ and $\overline{B} = \mathcal{P}(B) \backslash \mathcal{B}$. The separator $S = A \cap B$ is complete so that the marginal distribution of Y_S belongs to the saturated model. The asymptotic variance of the ML estimate of λ^S follows thus by Theorem 4.21 and has form,

$$\text{avar}(\hat{\lambda}_{\circledast}^S) = \left(M_S^T \mathbb{P}^{-S} M_S \right)_{\circledast}. \qquad (4.50)$$

If λ is the log-linear parameter of the distribution of Y_V under $M(G)$, then $\lambda = (\lambda_V, \lambda_{\overline{V}})$ where $\lambda_{\overline{V}} = 0$, and it is shown in equation (4.21) that $\lambda_V = (\lambda_{A\backslash B}, \lambda_S, \lambda_{B\backslash A})$ where

$$\lambda_{A\backslash B} = \lambda^A_{A\backslash B} \text{ and } \lambda_{B\backslash A} = \lambda^B_{B\backslash A}.$$

Furthermore, because (A, B) forms a decomposition of G, it follows from Corollary 4.11 that

$$\hat{\lambda}_{A\backslash B} = \hat{\lambda}^A_{A\backslash B} \text{ and } \hat{\lambda}_{B\backslash A} = \hat{\lambda}^B_{B\backslash A}.$$

The following lemma shows that $\hat{\lambda}_{A\backslash B}$ and $\hat{\lambda}_{B\backslash A}$ are asymptotically uncorrelated, and therefore independent in their joint asymptotic normal distribution.

Lemma 4.24 *Let λ be the log-linear parameter of the cross-classified Bernoulli random vector Y_V with probability parameter $\pi > 0$. Furthermore, let $G = (V, E^\sim)$ be an undirected graph and assume that the probability distribution of Y_V belongs to $M(G)$. If (A, B) forms a decomposition of G then*

$$\text{avar}(\hat{\lambda}_{\mathscr{e}})_{A\backslash B, B\backslash A} = 0.$$

Proof. We have to show that

$$\text{avar}(\hat{\lambda}_{\mathscr{e}})_{A\backslash B, B\backslash A} = \frac{\partial \lambda_{A\backslash B}}{\partial \mu^T_{B\backslash A}} = 0.$$

where $\lambda_{A\backslash B} = \lambda^A_{A\backslash B}$ by Theorem 4.9. By Lemma 4.10, $\mu^T_{V_{\mathscr{e}}}$ can be partitioned into $(\mu^T_{A_{\mathscr{e}}}, \mu^T_{B\backslash A})$; recall that $\mu(\varnothing) = 1$. As (A, B) is a decomposition of G, then the graphical model with graph G is collapsible onto A (Asmussen and Edwards, 1983) and this implies that the distribution of Y_A is Markov respect to $G_A = (A, E_A)$. The corresponding mean parameter is $\mu_{A_{\mathscr{e}}}$, whereas the canonical parameter is $\lambda^A_{A_{\mathscr{e}}}$. This implies that $\lambda^A_{A\backslash B}$ can be written as a function of $\mu_{A_{\mathscr{e}}}$ and, in turn, that we can apply the chain rule to obtain

$$\frac{\partial \lambda_{A\backslash B}}{\partial \mu^T_{B\backslash A}} = \frac{\partial \lambda^A_{A\backslash B}}{\partial \mu^T_{B\backslash A}} = \frac{\partial \lambda^A_{A\backslash B}}{\partial \mu^T_{A_{\mathscr{e}}}} \times \frac{\partial \mu_{A_{\mathscr{e}}}}{\partial \mu^T_{B\backslash A}}.$$

The vector $\mu_{V_{\otimes}}$ is the mean parameter of a regular exponential family in a full representation which take values in its domain $\mu(\Theta_{V_{\otimes}})$ and there are no functional relationships between its elements. Hence, because the subvectors $\mu_{A_{\otimes}}$ and $\mu_{B\backslash A}$ have no common elements, it holds that $\partial\mu_{A_{\otimes}}/\partial\mu_{B\backslash A}^{T} = 0$ which implies the desired result. □

We remark that Lemma 4.24 is a consequence of the connection between decomposition of the graph and cut in Y_V; see Lauritzen (1996, p. 242) and Piccioni (2000). Furthermore, for the decomposable case this result is implicit in the theory of meta Markov models introduced by Dawid and Lauritzen (1993, section 4); see also Consonni and Leucari (2006, Proposition 4).

As a consequence of Lemma 4.24 we obtain the local computation of the asymptotic variance of $\hat{\lambda}_V$.

Theorem 4.25 *In the setting of Lemma 4.24 it holds that*

$$\text{avar}(\hat{\lambda}_{V_{\otimes}}) = [\text{avar}(\hat{\lambda}_{A_{\otimes}}^A)]^0 + [\text{avar}(\hat{\lambda}_{B_{\otimes}}^B)]^0 - [\text{avar}(\hat{\lambda}_{\otimes}^S)]^0$$

and

$$|\text{avar}(\hat{\lambda}_{V_{\otimes}})| = \frac{|\text{avar}(\hat{\lambda}_{A_{\otimes}}^A)||\text{avar}(\hat{\lambda}_{B_{\otimes}}^B)|}{|\text{avar}(\hat{\lambda}_{\otimes}^S)|}$$

where $S = A \cap B$ *and for* $Q \subseteq V$, $[\text{avar}(\hat{\lambda}_{Q_{\otimes}}^Q)]^0$ *denotes the matrix indexed by* $V_{\otimes} \times V_{\otimes}$ *obtained by padding zero entries to* $\text{avar}(\hat{\lambda}_{Q_{\otimes}}^Q)$ *to obtain full dimension. Explicit forms for* $\text{avar}(\hat{\lambda}_{A_{\otimes}}^A)$ *and* $\text{avar}(\hat{\lambda}_{B_{\otimes}}^B)$ *are given in (4.49) and (4.49a), whereas for* $\text{avar}(\hat{\lambda}_{\otimes}^S)$ *see (4.50).*

Proof. By Lemma 4.10, $(A\backslash B, S_{\otimes}, B\backslash A)$ forms a partition of $V_{\otimes}$ and this, in turn, induces a block partition in $\text{avar}(\hat{\lambda}_{V_{\otimes}})$ where the block $\text{avar}(\hat{\lambda}_{\otimes})_{A\backslash B,B\backslash A}$ is equal to zero by Lemma 4.24 and we can apply Lauritzen (1996, Lemma 5.5) to write

$$\text{avar}(\hat{\lambda}_{V_{\otimes}}) = V(\mu_{V_{\otimes}})^{-1} = \left[\{V(\mu_{V_{\otimes}})_{A_{\otimes}}\}^{-1}\right]^0 + \left[\{V(\mu_{V_{\otimes}})_{B_{\otimes}}\}^{-1}\right]^0$$
$$- \left[\{V(\mu_{V_{\otimes}})_{S_{\otimes}}\}^{-1}\right]^0.$$

It follows from the definition of variance function that if $C \subseteq V$ is an expanded generating class then $V(\mu_{V_{\otimes}})_{C_{\otimes}} = V(\mu_{C_{\otimes}})$ and, therefore,

$$\mathrm{avar}(\hat{\lambda}_{V_\star}) = [V(\mu_{A_\star})^{-1}]^0 + [V(\mu_{B_\star})^{-1}]^0 - [V(\mu_{S_\star})^{-1}]^0$$

and the result follows because $V(\mu_{A_\star})^{-1} = \mathrm{avar}(\hat{\lambda}^A_{A_\star})$, $V(\mu_{B_\star})^{-1} = \mathrm{avar}(\hat{\lambda}^B_{B_\star})$ and $V(\mu_{S_\star})^{-1} = \mathrm{avar}(\hat{\lambda}^S_{S_\star})$. The factorization of the determinant can be shown in the same way. $\qquad\square$

Finally, we consider the decomposable case and show that the asymptotic variance can be locally computed with respect to the saturated models corresponding to cliques and separators. Furthermore, in this case an explicit form for the entries of the asymptotic variance is available.

Theorem 4.26   *In the setting of Lemma 4.24, assume that G is a decomposable graph with cliques $C_1, \ldots, C_k$ and separators $S_2, \ldots, S_k$. Then*

$$\mathrm{avar}(\hat{\lambda}_{V_\star}) = \sum_{i=1}^{k}[\mathrm{avar}(\hat{\lambda}^{C_i}_\star)]^0 - \sum_{i=2}^{k}[\mathrm{avar}(\hat{\lambda}^{S_i}_\star)]^0$$

and the entry indexed by $D, D' \in V_\star$ has the form

$$\mathrm{avar}(\hat{\lambda}_\star)_{D,D'} = \sum_{E \subseteq D \cap D'} (-1)^{|D \backslash E| + |D' \backslash E|}$$

$$\left\{ \sum_{i=1}^{k} \frac{\mathbf{1}(D \cup D' \subseteq C_i)}{\pi^{C_i}(E)} - \sum_{i=2}^{k} \frac{\mathbf{1}(D \cup D' \subseteq S_i)}{\pi^{S_i}(E)} \right\}.$$

Furthermore,

$$|\mathrm{avar}(\hat{\lambda}_{V_\star})| = \frac{\prod_{i=1}^{k}|\mathrm{avar}(\hat{\lambda}^{C_i}_\star)|}{\prod_{i=2}^{k}|\mathrm{avar}(\hat{\lambda}^{S_i}_\star)|} = \frac{\prod_{i=2}^{k}|\mathbb{P}^{S_i}|}{\prod_{i=1}^{k}|\mathbb{P}^{C_i}|}.$$

Proof.   The formulae for $\mathrm{avar}(\hat{\lambda}_{V_\star})$ and $|\mathrm{avar}(\hat{\lambda}_{V_\star})|$ follow from the recursive application of Theorem 4.25 with respect to a perfect sequence of cliques of G and then by applying (4.43).

Consider now the entries of $\mathrm{avar}(\hat{\lambda}_{V_\star})$. If A is a complete subset of V, then the entry of $[\mathrm{avar}(\hat{\lambda}^A_\star)]^0$ indexed by $D, D' \in V_\star$ is given in Theorem 4.21 if $D, D' \subseteq A$ and is equal to zero otherwise. Because $D, D' \subseteq A$ if and only if $\mathbf{1}(D \cup D' \subseteq A) = 1$ we can write

$$[\mathrm{avar}(\hat{\lambda}^A_\star)]^0_{D,D'} = \sum_{E \subseteq D \cap D'} (-1)^{|D \backslash E| + |D' \backslash E|} \frac{\mathbf{1}(D \cup D' \subseteq A)}{\pi^A(E)} \qquad (4.51)$$

and the expression for the entries of $\text{avar}(\hat{\lambda}_{V_s})$ can be obtained by applying (4.51) with respect to the cliques and separators. $\qquad\square$

We close this section by noticing that the factorization of the determinant in Theorem 4.25 implies a corresponding factorizations of the determinant of the variance function because $|V(\mu_{V_s})| = |\text{avar}(\hat{\lambda}_{V_s})|^{-1}$, $|V(\mu_{A_s})| = |\text{avar}(\hat{\lambda}^A_{A_s})|^{-1}$, $|V(\mu_{B_s})| = |\text{avar}(\hat{\lambda}^B_{B_s})|^{-1}$ and, finally, $|V(\mu_{S_s})| = |\text{avar}(\hat{\lambda}^S_s)|^{-1}$. Clearly, a similar result also holds for the factorization of the determinant in Theorem 4.26.

5

Bidirected Graph Models

Undirected graph models put all the variables of the relevant random vector Y_V on an equal standing. Accordingly, the edges of the graph are of a symmetric type to encode a likewise symmetric relationship between variables. When an edge is missing the corresponding variables are conditionally independent given all the remaining variables in Y_V. More generally, each of the three Markov properties for undirected graphs is characterized by a collection of conditional independence relationships of the type $Y_A \perp\!\!\!\perp Y_B | Y_{V \setminus (A \cup B)}$, for suitable subsets A and B of V. In other words, undirected graph models are defined through independence relationships fulfilled by conditional distributions involving all the variables in Y_V, and in this sense we can say that they are "conditional" independence models. This chapter is devoted to models defined by bidirected graphs, i.e., graphs whose edges are arrows with two heads ($\leftrightarrow$). Like undirected edges, bidirected edges encode symmetric relationships between variables and, in the corresponding independence models, the variables are considered to be on an equal standing. The distinguishing feature of these models is that every missing edge implies the marginal independence of the corresponding variables, rather than the conditional independence. There are three distinct Markov properties for bidirected graphs, and each is characterized by a collection of marginal independencies of the type $Y_A \perp\!\!\!\perp Y_B$, where A and B are suitable subsets of V. For this reason, bidirected graph models can be regarded as "marginal" independence models. The interest for the marginal independence restrictions encoded by a bidirected graph is due to the fact that they may arise through confounding effects of unobserved variables (Cox and Wermuth, 1993, 1996; Richardson and Spirtes, 2002; Drton and Richardson, 2008a).

Graphical models of marginal independence were first introduced for Gaussian random vectors. Marginal independencies in Gaussian distributions correspond to zero covariances, and Cox and Wermuth (1993,

1996) used a *covariance graph* to represent the zero pattern of a covariance matrix. In this context a relevant role is also played by the seminal work by Anderson (1969, 1973) on models defined by linear restrictions on covariances. In a covariance graph, every missing edge implies the marginal independence of the corresponding variables. Models for categorical variables with analogous independence structure were first developed by Kauermann (1997), and a comprehensive set of references on this area can be found in Drton and Richardson (2008a) and La Rocca and Roverato (2017). Notice that dashed undirected lines are sometimes used in place of bidirected arrows in the graphical representation of these models. We adopt the latter convention because it is consistent with the notation used in the path diagrams of Wright (1921).

In the following, we will present the theory concerning the Markov properties for bidirected graphs and the relative statistical models for categorical variables. We will deal with inferential issues in the next chapter. The latter gives the theory of regression graph models which include models for bidirected graphs as a special case.

5.1 Bidirected Graphs

We provide here some additional notions of graph theory required to deal with models defined by bidirected graphs.

A *bidirected graph* $G = (V, E)$ is a graph with only bidirected edges, and when we want to highlight that a graph is bidirected we write its edge set as $E^{\leftrightarrow}$. Figure 5.1 gives an example of bidirected graph that is the bidirected version of the graph in Figure 4.1. Several notions introduced for undirected graphs can be defined in the same way for bidirected graphs, and we will apply to bidirected graphs the concepts of *induced subgraph, path* and *separation*. Hence, similarly to what happens in the graph of Figure 4.1, also in the graph of Figure 5.1 the

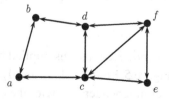

Figure 5.1 Bidirected version of the graph in Figure 4.1.

sequence (a, b, d, c, f) is a path from a to f. Furthermore, the subsets $\{c, d\}$ and $\{b, c\}$ both separate the vertex a from the vertex f, and also the subset $A = \{a\}$ from the subset $B = \{e, f\}$. We will also apply to bidirected graphs the notion of *neighbor* that, for a subset $A \subseteq V$, is denoted by $\mathrm{nb}(A)$ and defined just as in the undirected case. Although we do not use it here, we remark that in the literature of bidirected graph models the notion of neighbor is often replaced by that of *spouse*. In our notation, the spouse of a subset $A \subseteq V$, as defined, e.g., in Drton and Richardson (2008a), is the set of vertices $A \cup \mathrm{nb}(A)$.

A key role in the theory of bidirected graph models is played by the connected subsets of the vertex set V. A graph $G = (V, E)$ is *connected* when every pair of distinct vertices in V is joined by a path. A nonempty subset $A \subseteq V$ is a *connected set* in G if the induced subgraph G_A is connected, and it is *disconnected* otherwise. Every nonempty subset $A \subseteq V$ can be partitioned uniquely into maximal connected sets, $A = C_1 \cup \ldots \cup C_r$. The sets $C_1, \ldots, C_r$ are called the *connected components* of A. For example, in the graph of Figure 5.1 the set $\{a, b, c, e, f\}$ is connected, whereas the set $\{a, b, e, f\}$ is disconnected with connected components $C_1 = \{a, b\}$ and $C_2 = \{e, f\}$. Also the set $\{a, d, e\}$ is disconnected and its connected components are the singletons $C_1 = \{a\}$, $C_2 = \{d\}$ and $C_3 = \{e\}$.

5.2 Markov Properties for Bidirected Graphs

There are three distinct Markov properties for bidirected graphs, each being the dual of one of the Markov properties for undirected graphs (Kauermann, 1996).

The Dual Pairwise Markov Property

The probability distribution of a random vector Y_V is said to obey the *dual pairwise Markov property* with respect to the bidirected graph $G = (V, E^{\leftrightarrow})$ if for every $a, b \in V$ such that $\{a, b\} \notin E^{\leftrightarrow}$ it holds that

$$Y_a \perp\!\!\!\perp Y_b.$$

For example, two independence relations implied by this Markov property for the graph of Figure 5.1 are $Y_a \perp\!\!\!\perp Y_f$ and $Y_b \perp\!\!\!\perp Y_c$. It is useful to compare these marginal independencies with the corresponding conditional independencies implied by the pairwise Markov property for the graph of Figure 4.1. Duality of the two Markov properties is clear: every missing edge $a \neq b$ implies the independence relationship $Y_a \perp\!\!\!\perp Y_b | Y_C$ where in the undirected case the conditioning set is the largest possible, i.e., $C = V \backslash \{a, b\}$, whereas in the bidirected case it is the smallest possible, i.e., $C = \varnothing$.

The Dual Local Markov Property

The probability distribution of a random vector Y_V is said to obey the *dual local Markov property* with respect to the bidirected graph $G = (V, E^{\leftrightarrow})$ if for every vertex $a \in V$ it holds that

$$Y_a \perp\!\!\!\perp Y_{V \backslash (\mathrm{nb}(a) \cup \{a\})}.$$

For example, if the distribution of Y_V is locally Markov with respect to the graph of Figure 5.1 then it holds that $Y_a \perp\!\!\!\perp Y_{\{d,e,f\}}$, $Y_d \perp\!\!\!\perp Y_{\{a,e\}}$ and $Y_c \perp\!\!\!\perp Y_b$. Also in this case, the analogy with the local Markov property for undirected graphs is straightforward.

The Dual Global Markov Property

The probability distribution of a random vector Y_V is said to obey the *dual global Markov property* with respect to the bidirected graph $G = (V, E^{\leftrightarrow})$ if for any triple of pairwise disjoint subsets $A, B, S \subseteq V$ such that S separates A from B in G it holds that

$$Y_A \perp\!\!\!\perp Y_B | Y_{V \backslash (A \cup B \cup S)}.$$

For this Markov property, duality with the relative Markov property for undirected graphs is more subtle. Like in the undirected case, every separation implies a conditional independence relationship, but the conditioning set is the complement of the separating set S with respect to $V \backslash (A \cup B)$ rather than S itself. Examples of conditional independencies implied by the dual global Markov property for the graph of Figure 5.1 are $Y_a \perp\!\!\!\perp Y_{\{e,f\}} | Y_b$ and $Y_{\{b,d\}} \perp\!\!\!\perp Y_e | Y_a$.

The above definition of global Markov property for bidirected graph is useful because it provides a rule for reading a wide range of independence relationships from the graph, including conditional ones. More specifically, as shown in Richardson and Spirtes (2002), the dual global Markov property is complete; see also Drton and

Richardson (2008a). We recall that this means that the conditional independencies encoded by any bidirected graph G under the dual global Markov property are the only conditional independencies that are simultaneously satisfied by all the distributions which are (dual) globally Markov with respect to G. On the other hand, this Markov property has the disadvantage of being formulated in terms of conditional independencies, that may hide its "marginal" nature. In this sense, of special interest is the case where $V \setminus (A \cup B)$ separates A from B in G because the dual global Markov property implies the marginal independence of Y_A and Y_B. As shown in Proposition 4.1, the global Markov property for undirected graphs is characterized by the independence relationships between subvectors Y_A and Y_B implied by separating sets of the type $V \setminus (A \cup B)$. This is also true for the dual global Markov property and it has led to the introduction of a further, but equivalent, Markov property based on marginal independencies.

The Connected Set Markov Property

For a bidirected graph $G = (V, E^{\leftrightarrow})$ consider two disjoint subsets of vertices A and B. It is not difficult to see that the subset $V \setminus (A \cup B)$ separates A from B in G if and only if there is no edge joining a vertex in A with a vertex in B, that is, if and only if A and B are not connected to each other in $G_{A \cup B}$. In this case the set $A \cup B$ is disconnected in G and the dual global Markov property implies $Y_A \perp\!\!\!\perp Y_B$. This observation suggested the introduction of a Markov property that focuses on the connected subsets of V. Formally, the probability distribution of a random vector Y_V is said to obey the *connected set Markov property* with respect to bidirected graph $G = (V, E^{\leftrightarrow})$ if for every disconnected set A of G it holds that

$$Y_{C_1} \perp\!\!\!\perp \ldots \perp\!\!\!\perp Y_{C_r},$$

where $C_1, \ldots, C_r$ are the connected components of A. For example, in the graph of Figure 5.1 the subset $\{a, b, e, f\}$ is disconnected with connected components $\{a, b\}$ and $\{e, f\}$ and therefore under the connected set Markov property it holds that $Y_{\{a,b\}} \perp\!\!\!\perp Y_{\{e,f\}}$. In the same graph, the subset $\{a, d, e\}$ is disconnected with connected components $\{a\}$, $\{d\}$ and $\{e\}$ and therefore the connected set Markov property implies $Y_a \perp\!\!\!\perp Y_d \perp\!\!\!\perp Y_e$.

The connected set Markov property is equivalent to the dual global Markov property as formally stated by the following proposition proved in Richardson (2003); see also Drton and Richardson (2008a).

Proposition 5.1 *The probability distribution of Y_V obeys the dual global Markov property with respect to a bidirected graph $G = (V, E^{\leftrightarrow})$ if and only if it obeys the connected set Markov property with respect to G.*

Proof. Assume that the distribution of Y is dual globally Markov with respect to $G = (V, E^{\leftrightarrow})$, and let $A \subseteq V$ be an arbitrary disconnected set with connected components $C_1, \ldots, C_r$. If we let $A_2 = C_2 \cup \ldots \cup C_r$, then $V \backslash A$ separates C_1 from A_2 in G, and therefore by the dual global Markov property $Y_{C_1} \perp\!\!\!\perp Y_{A_2}$. By iteratively applying the same procedure to the pairs of subsets C_i and $A_{i+1} = C_{i+1} \cup \ldots \cup C_r$, for $i = 1, \ldots, (r-1)$, one obtains $Y_{C_1} \perp\!\!\!\perp \ldots \perp\!\!\!\perp Y_{C_r}$, and therefore that Y obeys the connected set Markov property with respect to G. Consider now the case where the distribution of Y obeys the connected set Markov property with respect to G and assume that A, B and S is an arbitrary triple of pairwise disjoint subsets of V such that S separates $A \neq \emptyset$ form $B \neq \emptyset$ in G. Then, one can always find two disjoint sets D_1 and D_2 such that $D_1 \cup D_2 = V \backslash (A \cup B \cup S)$ and S separates $A \cup D_1$ from $B \cup D_2$ in G. Hence, the sets $A \cup D_1$ and $B \cup D_2$ are not connected to each other in G and the connected set Markov property implies that $Y_{A \cup D_1} \perp\!\!\!\perp Y_{B \cup D_2}$. In turn, it follows from the properties of conditional independence (see Lauritzen, 1996, section 3.1) that $Y_A \perp\!\!\!\perp Y_B | Y_{D_1 \cup D_2}$, i.e., that $Y_A \perp\!\!\!\perp Y_B | Y_{V \backslash (A \cup B \cup S)}$ as required. □

The connected set Markov property is not complete, and therefore it is less exhaustive than the dual global Markov property. However, it makes it clear that every independence model associated with a bidirected graph is characterized by a collection of marginal independencies. Furthermore, the connected set Markov property is useful in the specification of suitable parameterizations of bidirected graph models. Indeed, unlike the undirected case where for positive distributions one can exploit the equivalence of the three Markov properties, this is not possible in the bidirected case. To see this, we now turn to the analysis of the connections among the four Markov properties for bidirected graphs. It is easy to see that the dual global Markov property implies the dual local Markov property because for every $a \in V$ it holds that

Table 5.1 Probability distribution for a binary random vector $Y_{\{a,b,c\}}$ such that both $Y_a \perp\!\!\!\perp Y_b$ and $Y_a \perp\!\!\!\perp Y_c$ but $Y_a \not\perp\!\!\!\perp Y_{\{b,c\}}$.

		Y_c	
Y_a	Y_b	0	1
0	0	0.02	0.08
	1	0.03	0.12
1	0	0.05	0.25
	1	0.10	0.35

$\mathrm{nb}(a)$ separates a from $V \backslash (\mathrm{nb}(a) \cup \{a\})$ in G. Furthermore, the dual local Markov property implies the dual pairwise Markov property because if $\{a, b\} \notin E^{\leftrightarrow}$ then $b \in V \backslash (\mathrm{nb}(a) \cup \{a\})$, and therefore the local independence $Y_a \perp\!\!\!\perp Y_{V \backslash (\mathrm{nb}(a) \cup \{a\})}$ implies the pairwise independence $Y_a \perp\!\!\!\perp Y_b$. We can thus summarize the relationships between the Markov properties for bidirected graphs as follows,

(connected set) $\Leftrightarrow$ (dual global) $\Rightarrow$ (dual local) $\Rightarrow$ (dual pairwise).

Unlike the undirected case, the dual pairwise Markov property does not imply the dual global Markov property, even for positive distributions. To show this, Table 5.1 provides an example, taken from Drton and Richardson (2008a), of a positive probability distribution for the binary vector $Y_{\{a,b,c\}}$ with both $Y_a \perp\!\!\!\perp Y_b$ and $Y_a \perp\!\!\!\perp Y_c$ but where Y_a is not independent of $Y_{\{b,c\}}$. Hence, this distribution obeys the dual pairwise Markov property with respect to the graph in Figure 5.2C but not the dual global Markov property with respect to the same graph.

5.3 The Log-mean Linear Parameterization

Marginal independencies between categorical variables cannot be conveniently implemented through the log-linear parameterization because marginal independence constraints do not correspond to linear restrictions in the log-linear parameter space. Nevertheless, a suitable parameterization can be obtained by exploiting the existing duality between the undirected and the bidirected case, which also extends to the parameterization level. The log-linear parameter λ is obtained from the log-

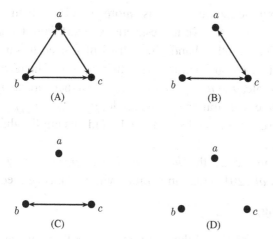

Figure 5.2 Four graphs of marginal independence for $Y_{\{a,\ b,\ c\}}$.

linear expansion of the probability parameter π, and the role it plays in undirected graph models follows from the existing connection between its terms and cross-product ratios of the "conditional" type, $\mathrm{cpr}_{cc}(\cdot)$. The cross-classified Bernoulli distribution belongs to the regular exponential family, and a parameterization alternative to the probability parameter π is provided by the mean parameter μ. This is a vector of marginal probabilities and its log-linear expansion yields a parameter γ named the *log-mean linear parameter* of Y. Hereafter, we will describe the connection existing between the entries of γ and the cross-product ratios of the "marginal" type, $\mathrm{cpr}_{mm}(\cdot)$, used to assess marginal independence relationships.

The Binary Case

For a binary random vector Y with probability parameter π the mean parameter, as defined in Section 4.10.4, is given by $\mu = (\mu_D)_{D \subseteq V}$ where

$$\mu_D = p(Y_D = 1) \quad \text{for every } D \subseteq V,$$

with the convention that $\mu_\emptyset = 1$. We recall that by Proposition 4.16 the vector μ can be computed as $\mu = \mathbb{Z}\pi$ so that $\pi = \mathbb{M}\mu$. As an example, we give below the mean parameter relative to Table 5.1,

$\mu_\emptyset$	μ_a	μ_b	μ_c	$\mu_{\{a,b\}}$	$\mu_{\{a,c\}}$	$\mu_{\{b,c\}}$	$\mu_{\{a,b,c\}}$
1	0.75	0.60	0.80	0.45	0.60	0.47	0.35

The parametric space of μ is more involved than that of π. The probabilities in π have no restrictions, apart from the sum-to-one constraint. On the other hand, the entries of μ need to satisfy several constraints. It is straightforward to see that $\mu_{D'} \geq \mu_D$ whenever $D' \subseteq D$ so that, for example, $\mu_a \geq \mu_{\{a,b\}}$ and $\mu_b \geq \mu_{\{a,b\}}$, but there are also additional, less-immediate, constraints. For example, $\mu_a + \mu_b - \mu_{\{a,b\}}$ is the probability of the event $\{Y_a = 1\} \cup \{Y_b = 1\}$ and this implies that both $\mu_a + \mu_b \geq \mu_{\{a,b\}}$ and $\mu_a + \mu_b \leq 1 + \mu_{\{a,b\}}$.

We now introduce the *log-mean linear parameter* $\gamma = (\gamma_D)_{D \subseteq V}$ (Roverato *et al.*, 2013) that, in parallel with (4.1), is defined as

$$\gamma = \mathbb{M}^T \log \mu. \tag{5.1}$$

It follows from (5.1) and the fact that $\mu = \mathbb{Z}\pi$ that one can compute γ from π as $\gamma = \mathbb{M}^T \log \mathbb{Z}\pi$, and thus π from γ as $\pi = \mathbb{M}\exp(\mathbb{Z}^T\gamma)$. Hence, the functional relationship between γ and π is a smooth bijection and this implies that γ is a valid parameterization of the distribution of Y.

The individual entries of γ can be written by (3.3) as

$$\gamma_D = \sum_{D' \subseteq D} (-1)^{|D \setminus D'|} \log \mu_{D'} \text{ for every } D \subseteq V, \tag{5.2}$$

and this also shows that the log-mean linear parameters satisfy a property called *upward compatibility*, i.e., every interaction term γ_D is computed in the distribution of the relevant margin Y_D. It is also useful to compare the theoretical quantities with a more concrete example and to this aim we provide the log-mean linear terms relative to Table 5.1,

$\gamma_\emptyset$	γ_a	γ_b	γ_c	$\gamma_{\{a,b\}}$	$\gamma_{\{a,c\}}$	$\gamma_{\{b,c\}}$	$\gamma_{\{a,b,c\}}$
0	−0.288	−0.511	−0.223	0	0	−0.021	−0.00712

The computation of the log-mean linear parameter is based on a Möbius inversion formula, and this allows us to exploit the tools introduced in Chapter 3. The rest of this section shows the connection between independence relationship and vanishing log-mean linear terms, and it runs in parallel with the theory of the log-linear parameterization given in Section 4.3.

The entry of γ indexed by the empty set is always equal to zero,

$$\gamma_\emptyset = \log \mu_\emptyset = \log(1) = 0.$$

The main effect γ_a, that is the log-mean linear term for the singleton set $\{a\}$, is computed in the marginal distribution of Y_a,

$$\gamma_a = \log \frac{\mu_a}{\mu_\emptyset} = \log \mu_a = \log p(Y_a = 1).$$

The two-way interaction relative to $D = \{a, b\}$ is computed in the marginal distribution of $Y_{\{a,b\}}$, and it is the logarithm of a cross-product ratio,

$$
\begin{aligned}
\gamma_{\{a,b\}} &= \log \frac{\mu_{\{a,b\}} \mu_\emptyset}{\mu_a \mu_b} \\
&= \log \frac{p(Y_a = 1, Y_b = 1)}{p(Y_a = 1) p(Y_b = 1)} \\
&= \log \mathrm{cpr}_{mm}(a, a : b, b).
\end{aligned}
$$

It follows from Lemma 2.2 that $\gamma_{\{a,b\}} = 0$ is equivalent to the independence of Y_a and Y_b. Similarly to the two-way interaction $\gamma_{\{a,b\}}$, the term $\gamma_{\{a,c\}}$ is

$$\gamma_{\{a,c\}} = \log \frac{\mu_{\{a,c\}} \mu_\emptyset}{\mu_a \mu_c} = \log \mathrm{cpr}_{mm}(a, a : c, c),$$

and therefore $Y_a \perp\!\!\!\perp Y_c$ if and only if $\gamma_{\{a,c\}} = 0$. Consider for example the log-mean linear terms given above, relative to Table 5.1. For this probability distribution $\gamma_{\{a,b\}} = \gamma_{\{a,c\}} = 0$ and therefore it holds both that $Y_a \perp\!\!\!\perp Y_b$ and $Y_a \perp\!\!\!\perp Y_c$.

The three-way interaction for $\{a, b, c\}$ can be decomposed as

$$
\begin{aligned}
\gamma_{\{a,b,c\}} &= \log \frac{\mu_{\{a,b,c\}} \, \mu_a \, \mu_b \, \mu_c}{\mu_{\{a,b\}} \, \mu_{\{a,c\}} \, \mu_{\{b,c\}} \, \mu_\emptyset} \\
&= \log \frac{\mu_{\{a,b,c\}}}{\mu_{\{a\}} \, \mu_{\{b,c\}}} - \log \frac{\mu_{\{a,b\}}}{\mu_{\{a\}} \, \mu_{\{b\}}} - \log \frac{\mu_{\{a,c\}}}{\mu_{\{a\}} \, \mu_{\{c\}}} \\
&= \log \mathrm{cpr}_{mm}(a, a : \{b, c\}, \{b, c\}) - \gamma_{\{a,b\}} - \gamma_{\{a,c\}}.
\end{aligned}
$$

By Lemma 2.2 it holds that $Y_a \perp\!\!\!\perp Y_{\{b,c\}}$ if and only if

$$
\begin{aligned}
\mathrm{cpr}_{mm}(a, a : \{b, c\}, \{b, c\}) &= \mathrm{cpr}_{mm}(a, a : b, \{b, c\}) \\
&= \mathrm{cpr}_{mm}(a, a : c, \{b, c\}) = 1
\end{aligned}
$$

where

$$\mathrm{cpr}_{mm}(a, a : b, \{b, c\}) = \mathrm{cpr}_{mm}(a, a : b, b) = \gamma_{\{a,b\}}$$

and

$$\mathrm{cpr}_{mm}(a, a : c, \{b, c\}) = \mathrm{cpr}_{mm}(a, a : c, c) = \gamma_{\{a,c\}}.$$

Hence, when $\gamma_{\{a,b\}} = \gamma_{\{a,c\}} = 0$, the vanishing of the three-way interaction, $\gamma_{\{a,b,c\}} = 0$, is a necessary and sufficient condition for the independence of Y_a and $Y_{\{b,c\}}$. In the example of Table 5.1 we have $\gamma_{\{a,b,c\}} = -0.00712$ and thus $Y_a \not\perp\!\!\!\perp Y_{\{b,c\}}$.

We can now formally state the connection between the vanishing of certain log-mean linear interactions and independence relationships. This theorem is the dual of the equivalent result for the undirected case given in Theorem 4.2. Also in this case the proof is an immediate consequence of the basic lemmas of Chapter 2, connecting cross-product ratios with independence relationships, together with Lemma 3.3 for Möbius inversion; see also Roverato *et al.* (2013) and La Rocca and Roverato (2017).

Theorem 5.2　　*For a vector Y_V of binary variables with probability parameter $\pi > 0$ let $\gamma = \mathbb{M}^T \log \mu$ where $\mu = \mathbb{Z}\pi$. Then, for a pair of disjoint nonempty subsets A and B of V the following conditions are equivalent:*

(i) $Y_A \perp\!\!\!\perp Y_B$;
(ii) *for every $D \subseteq A \cup B$ such that both $D \cap A \neq \emptyset$ and $D \cap B \neq \emptyset$ it holds that*

$$\gamma_D = 0;$$

(iii) *for every $A' \subseteq A$, $B' \subseteq B$, such that $A', B' \neq \emptyset$ it holds that*

$$\log \mu_{A' \cup B'} - \log \mu_{A'} - \log \mu_{B'} + \log \mu_\emptyset = 0. \qquad (5.3)$$

Proof. The equivalence of (i) and (iii) follows from Lemma 2.2 because the identity (5.3) is equal to $\log \mathrm{cpr}_{mm}(A', A : B', B) = 0$. The equivalence of (ii) and (iii) is that stated in Lemma 3.3 when $\theta = \gamma$ and $\omega = \log \mu$. □

The Non-binary Case

The extension of the parameterization μ to categorical variables with arbitrary number of levels was developed by Drton (2009). Let Y be a vector of categorical variables with state space $\mathcal{I}$ and restricted state space $\mathcal{J}$. Drton (2009) introduced the *saturated Möbius parameter*

$$\mu_j = (\mu_{j_D})_{D \subseteq V} \text{ for all } j \in \mathcal{J},$$

where

$$\mu_{j_D} = p(Y_D = j_D),$$

and showed that it smoothly parameterizes the distribution of Y. Consequently, the log-mean linear parameter γ of Y can be defined as the collection of vectors

$$\gamma_j = (\gamma_{j_D})_{D \subseteq V} \text{ for all } j \in \mathcal{J},$$

where

$$\gamma_j = \mathbf{M}^T \log \mu_j \text{ and } \gamma_{j_D} = (\gamma_j)_D.$$

Recall that we adopt the convention that $j_\emptyset = \mathcal{J}_\emptyset = \emptyset$. As well as for the probability parameter and for the log-linear parameter also in this case it is possible a more compact representation of the Möbius parameter of Y as $\mu = (\mu_{j_D})_{D \subseteq V, j_D \in \mathcal{J}_D}$. Likewise, $\gamma = (\gamma_{j_D})_{D \subseteq V, j_D \in \mathcal{J}_D}$.

A formal extension of Theorem 5.2 to the general case of categorical variables with arbitrary number of levels can be stated as follows; see also Roverato (2015) and La Rocca and Roverato (2017).

Corollary 5.3 *Let Y_V be a vector of categorical variables with probability parameter $\pi > 0$, Möbius parameter $\mu = (\mu_{j_D})_{D \subseteq V, j_D \in \mathcal{J}_D}$ and log-mean linear parameter $\gamma = (\gamma_{j_D})_{D \subseteq V, j_D \in \mathcal{J}_D}$. Then, for a pair of disjoint nonempty subsets A and B of V the following conditions are equivalent:*

(i) $Y_A \perp\!\!\!\perp Y_B$;

(ii) *for every $D \subseteq A \cup B$ such that both $D \cap A \neq \emptyset$ and $D \cap B \neq \emptyset$ it holds that*

$$\gamma_{j_D} = 0,$$

for every $j \in \mathcal{J}$;

(iii) *for every $A' \subseteq A, B' \subseteq B$, such that $A', B' \neq \emptyset$ it holds that*

$$\log \mu_{j_{A' \cup B'}} + \log \mu_{j_{A'}} + \log \mu_{j_{B'}} - \log \mu_\emptyset = 0, \tag{5.4}$$

for every $j \in \mathcal{J}$.

Proof. The equivalence of (i) and (iii) follows from Lemma 2.6 because the identity (5.4) is equal to $\log \mathrm{cpr}_{mm}(j_{A'}, A : j_{B'}, B) = 0$. The equivalence of (ii) and (iii) holds for every $j \in \mathcal{J}$ and is that stated in Lemma 3.3 when $\theta = \gamma_j$ and $\omega = \log \mu_j$.   □

5.4   Log-mean Linear Graphical Models

A graphical model is a family of probability distributions defined through the independence relationships encoded by a graph under a given Markov property. Unlike the undirected case where, for positive probability distributions, the graphical models under the three Markov properties coincide, in models defined by bidirected graphs it is always necessary to specify the Markov property being used. The results given in the previous section show that models for bidirected graphs can be identified by means of vanishing log-mean linear interactions. Specifically, Theorem 5.2 and Corollary 5.3 can be immediately applied to identify the log-mean linear restrictions required to specify a bidirected graph model under either the dual pairwise or the dual local Markov properties. Establishing the connection of the dual global Markov property with the log-mean linear parameter is less straightforward and carefully described below; pairwise independencies are also discussed for a comparison.

The Binary Case

The independence of two binary variables, say Y_a and Y_b, is equivalent to the vanishing of a single log-mean linear interaction, regardless of the dimension of Y_V. This follows from Theorem 5.2 where it is shown that Y_a and Y_b are independent if and only if $\gamma_{\{a,b\}} = 0$. It is thus immediate to assess whether a distribution belongs to a bidirected graph model under the dual pairwise Markov property because this happens whenever every two-way log-mean linear interaction associated with a missing edge is equal to zero. More formally, the distribution of Y_V obeys the dual pairwise Markov property with respect to $G = (V, E^{\leftrightarrow})$ if and only if for every pair of distinct vertices $a, b \in V$ such that $\{a, b\} \notin E^{\leftrightarrow}$ it holds that $\gamma_{\{a,b\}} = 0$.

The log-mean linear expansion of the individual entries of μ is

$$\log \mu_D = \sum_{D' \subseteq D} \gamma_{D'} \quad \text{for every } D \subseteq V,$$

so that the expansion of $\log \mu_V$ involves all the entries of γ. Consider the example where $V = \{a, b, c\}$. The saturated model is associated with the complete graph of Figure 5.2A, and this corresponds to the log-mean linear model with no restrictions, but for $\gamma_\emptyset = 0$. Hence we can write

$$\log \mu_{\{a,b,c\}} = \gamma_a + \gamma_b + \gamma_c + \gamma_{\{a,b\}} + \gamma_{\{a,c\}} + \gamma_{\{b,c\}} + \gamma_{\{a,b,c\}}. \quad (5.5)$$

To gain insight into the meaning of the log-mean linear expansion it is useful to notice that the upward compatibility property implies invariance of the log-mean linear terms with respect to marginalization. Thus, marginalization over some variables can be carried out by removing the proper log-mean linear terms. For instance, for the log-mean linear expansion of the probability distribution of $Y_{\{a,b\}}$ it holds that $\log \mu_{\{a,b\}} = \gamma_a + \gamma_b + \gamma_{\{a,b\}}$, where the γ-terms are the same as those in (5.5). By further removing the terms γ_b and $\gamma_{\{a,b\}}$ one obtains $\log \mu_a = \gamma_a$ that is the log-mean linear expansion relative to Y_a.

Under the dual pairwise Markov property, the model defined by the bidirected graph in Figure 5.2B is the log-mean linear model satisfying the constraint $\gamma_{\{a,b\}} = 0$,

$$\log \mu_{\{a,b,c\}} = \gamma_a + \gamma_b + \gamma_c + \gamma_{\{a,c\}} + \gamma_{\{b,c\}} + \gamma_{\{a,b,c\}}.$$

Accordingly, the distribution of Y belongs to the (pairwise) bidirected graph model for the graph Figure 5.2C if and only if both $\gamma_{\{a,b\}}$ and $\gamma_{\{a,c\}}$ are constrained to vanish,

$$\log \mu_{\{a,b,c\}} = \gamma_a + \gamma_b + \gamma_c + \gamma_{\{b,c\}} + \gamma_{\{a,b,c\}}.$$

Finally, the distribution of Y is pairwise Markov with respect to the graph in Figure 5.2D if and only if $\gamma_{\{a,b\}} = \gamma_{\{a,c\}} = \gamma_{\{b,c\}} = 0$, so that

$$\log \mu_{\{a,b,c\}} = \gamma_a + \gamma_b + \gamma_c + \gamma_{\{a,b,c\}}.$$

We now turn to the dual global Markov property. Consider the pairwise and the global bidirected graph models for the four graphs in Figure 5.2. For each of the graphs in Figures 5.2A and 5.2B the pairwise and the global Markov properties are equivalent. On the other hand, for the graph in Figures 5.2C the pairwise Markov property is different from the global Markov property, and the same is true for the graph in Figure 5.2D. This can be seen by noticing that in both graphs the subset $\{a, b, c\}$ is disconnected in such a way that the dual global Markov property implies $Y_a \perp\!\!\!\perp Y_{\{b,c\}}$. However, this independence is not satisfied by any of the respective pairwise models because the interaction $\gamma_{\{a,b,c\}}$ is unconstrained. This suggests that the disconnected subsets of V can be used to identify the vanishing log-mean linear interactions. The connection between zero restrictions in γ and the dual global Markov property can be formally proved by applying Theorem 5.2 to the independencies implied by the connected set Markov property (Roverato *et al.*, 2013, theorem 2).

Proposition 5.4   *Let Y_V be a binary random vector with probability parameter $\pi > 0$ and let $\gamma = \mathbb{M}^T \log \mu$ where $\mu = \mathbb{Z}\pi$. The distribution of Y_V obeys the connected set Markov property with respect to a bidirected graph G if and only if $\gamma_D = 0$ for every subset $D \subseteq V$ that is disconnected in G.*

Consider the graph in Figure 5.2C. The connected subsets of V are $\{a\}$, $\{b\}$, $\{c\}$ and $\{b, c\}$, whereas $\{a, b\}$, $\{a, c\}$ and $\{a, b, c\}$ are disconnected. Accordingly, the distribution of Y is globally Markov with respect to this graph if and only if $\gamma_{\{a,b\}} = \gamma_{\{a,c\}} = \gamma_{\{a,b,c\}} = 0$.

It is interesting to compare the dimensions of models for undirected graphs with those of models for bidirected graphs. Let $G_1 = (V, E^\sim)$ and $G_2 = (V, E^\leftrightarrow)$ be an undirected and a bidirected graph, respectively, with the same skeleton, i.e. with the same missing edges. An instance of this situation is provided by the graphs in Figures 4.1 and 5.1. In the distributions which are globally Markov with respect to G_1 the unrestricted log-linear terms are those indexed by the complete subsets of V. On the other hand, in the distributions which are dual globally Markov with respect to G_2 the unrestricted log-mean linear terms are those indexed by the connected subsets of V. Because every complete subset is connected, it follows that the undirected graph model is typically defined by a smaller number of parameters than the relative bidirected graph model. Thus, undirected graph models are typically more parsimonious than bidirected graph models. In turn, the model defined by a bidirected graph under the global Markov property is typically more parsimonious than the model defined by the same graph under the pairwise Markov property.

The Non-binary Case

The use of log-mean linear models to parameterize bidirected graph models can be straightforwardly extended to categorical variables with arbitrary number of levels as described in Roverato (2015). In this case the log-mean linear expansion of the Möbius parameters is

$$\log \mu_{j_D} = \sum_{D' \subseteq D} \gamma_{j_{D'}} \text{ for every } D \subseteq V \text{ and } j \in \mathcal{J},$$

and the constraint $\gamma_D = 0$ in the binary case needs to be reformulated as $\gamma_{j_D} = 0$ for every $j \in \mathcal{J}$. The connection between vanishing log-linear parameters and marginal independence follows from Corollary 5.3, which generalizes Theorem 5.2. This implies that, in the non-binary case, the distribution of Y obeys the dual pairwise Markov property with respect to $G = (V, E^{\leftrightarrow})$ if and only if for every pair of distinct vertices $a, b \in V$ such that $\{a, b\} \notin E^{\leftrightarrow}$ it holds that $\gamma_{j_{\{a,b\}}} = 0$ for every $j \in \mathcal{J}$. Furthermore, the distribution of Y obeys the connected set Markov property with respect to a bidirected graph G if and only if for every subset $D \subseteq V$ that is disconnected in G it holds that $\gamma_{j_D} = 0$ for every $j \in \mathcal{J}$ (Roverato, 2015, corollary 1).

5.5 Example: Symptoms in Psychiatric Patients

Table 5.2 shows data from Coppen (1966) for a set of four binary variables concerning symptoms in 362 psychiatric patients. The symptoms are: $Y_1 =$ stability (levels: extroverted, introverted), $Y_2 =$ validity (levels: psychasthenic, energetic), $Y_3 =$ acute depression (levels: yes, no) and $Y_4 =$ solidity (levels: hysteric, rigid).

In the graphical model literature, these data have been analyzed in two different ways. Firstly, Wermuth (1976) applied a conditional independence approach and selected the undirected graph model with

Table 5.2 Data from Coppen (1966) on symptoms in psychiatric patients.

		Solidity	Hysteric		Rigid	
Stability	**Depression**	**Validity**	**Psych.**	**Energetic**	**Psych.**	**Energetic**
Extrov.	No		12	47	8	14
	Yes		16	14	22	23
Introv.	No		27	46	22	25
	Yes		32	9	30	15

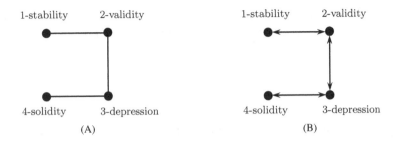

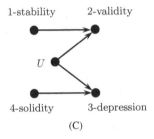

Figure 5.3 Graphs for Coppen data: (A) undirected graph, (B) bidirected graph, (C) directed graph including a vertex U for an unobserved variable.

graph in Figure 5.3A. Subsequently, Lupparelli *et al.* (2009) considered a marginal independence approach and obtained the bidirected graph model, under the dual global Markov property, for the graph in Figure 5.3B; see also Roverato *et al.* (2013). In applied work, the choice of the approach to follow usually depends on subject matter working hypotheses and background knowledge on the problem. This application to the Coppen data is of special interest because the two selected graphs have the same missing edges. It is thus of interest to compare the selected models with respect to their interpretation and statistical properties. Hereafter, we will denote the undirected and bidirected graphs in Figure 5.3 by $G^\sim = (V, E^\sim)$ and $G^\leftrightarrow = (V, E^\leftrightarrow)$, respectively, where $V = \{1, 2, 3, 4\}$.

The undirected graph $G^\sim$ is decomposable with cliques $\mathcal{C} = \{\{1, 2\}, \{2, 3\}, \{3, 4\}\}$ and separators $\{\{2\}, \{3\}\}$. For decomposable models a closed-form expression for the ML estimates is available, and the application of Proposition 4.12 shows that the ML estimates of the expected cell counts of the model for $G^\sim$ can be computed from the marginal observed counts as follows,

$$\hat{m}(Y = i) = \frac{n(Y_{\{1,2\}} = i_{\{1,2\}})n(Y_{\{2,3\}} = i_{\{2,3\}})n(Y_{\{3,4\}} = i_{\{3,4\}})}{n(Y_2 = i_2)n(Y_3 = i_3)}$$

for all $i \in \mathcal{I}$.

The dimension of the saturated model is $2^{|V|} - 1 = 15$, whereas the selected undirected graph model has dimension 7, with eight vanishing log-linear interactions,

$$\log \pi_{\{1,2,3,4\}} = \lambda_\varnothing + \lambda_1 + \lambda_2 + \lambda_3 + \lambda_4 + \lambda_{\{1,2\}} + \lambda_{\{2,3\}} + \lambda_{\{3,4\}}.$$

This selected model is characterized by the two conditional independencies $Y_1 \perp\!\!\!\perp Y_{\{3,4\}}|Y_2$ and $Y_4 \perp\!\!\!\perp Y_{\{1,2\}}|Y_3$, and it provides an adequate fit of the data with dev$(G^\sim) = 13.9$ on 8 degrees of freedom (p-value = 0.08). In order to interpret this model, we first recall that in undirected models every separation in the graph implies a connection between variables that is only indirect. This can be applied to see that, for example, depression (Y_3) separates validity (Y_2) from solidity (Y_4), and therefore the latter two variables are not associated directly, but only in an indirect way, through depression. Furthermore, for the prediction of a variable it is sufficient to have the information provided by its neighbors, in the sense that the remaining variables become irrelevant. Hence, for example, to predict solidity (Y_4) the variable depression (Y_3) is sufficient because the remaining two variables provide no additional information with respect to that provided by depression. Likewise, stability (Y_1) and depression (Y_3) are sufficient to predict validity (Y_2).

We now turn to the model for $G^{\leftrightarrow}$ under the dual global Markov property. The graph $G^{\leftrightarrow}$ has five disconnected sets, namely $\{1,3\}$, $\{1,4\}$, $\{2,4\}$, $\{1,2,4\}$ and $\{1,3,4\}$. Accordingly, the bidirected graph model defined by $G^{\leftrightarrow}$ has dimension 10 with five constrained log-mean linear interactions,

$$\log \mu_{\{1,2,3,4\}} = \gamma_1 + \gamma_2 + \gamma_3 + \gamma_4 + \gamma_{\{1,2\}} + \gamma_{\{2,3\}} + \gamma_{\{3,4\}} + \gamma_{\{1,2,3\}}$$
$$+ \gamma_{\{2,3,4\}} + \gamma_{\{1,2,3,4\}}.$$

As expected, this model is less parsimonious than the model defined by $G^\sim$. Inferential issues concerning bidirected graph models are deferred to the next chapter. Here we mention that the computation of the ML estimates for this model requires the application of an iterative procedure. Furthermore, this model also provides an adequate fit of the data (deviance equal to 8.6 on 5 degrees of freedom, p-value = 0.13).

The interpretation of the model implied by $G^{\leftrightarrow}$ is less intuitive than that implied by $G^{\sim}$. The bidirected graph model is characterized by the marginal independencies $Y_1 \perp\!\!\!\perp Y_{\{3,4\}}$ and $Y_4 \perp\!\!\!\perp Y_{\{1,2\}}$; that is, depression ($Y_3$) and solidity ($Y_4$) are jointly independent of stability (Y_1) and, moreover, stability (Y_1) and validity (Y_2) are jointly independent of solidity (Y_4). More interestingly, this model may suggest the presence of relevant unobserved variables. In the next chapter we will consider independence models defined by graphs with directed edges. We shall see that the graph in Figure 5.3C represents a multivariate regression model with three covariates, Y_1, Y_4 and U, and two responses, Y_2 and Y_3. The three covariates are mutually independent, and the responses are conditionally independent given the covariates. The system described by the graph in Figure 5.3C is thus clearly interpretable. When this system is marginalized over U, that is when U is unobserved, the independence structure of the four observed variables is that given by $G^{\leftrightarrow}$, and therefore the interpretation of the model for $G^{\leftrightarrow}$ is compatible with that of a multivariate regression model with unobserved variables.

5.6 Parsimonious Graphical Modeling

A general difficulty when dealing with categorical variables is that the dimension of the parametric space increases exponentially with the number of variables. Specifically, the distribution of a vector of $|V| = p$ binary variables has $2^p - 1$ parameters, and the number of parameters increases to $(\prod_{v \in V} |\mathcal{I}_v|) - 1$ in the polytomous case. For this reason, it is important to be able to implement additional substantive constraints so as to specify parsimonious submodels. Parsimonious modeling is an important issue in graphical modeling, and more widely in multivariate analysis, but it is a problem of especial relevance in bidirected graph modeling because the number of parameters can be relatively large even for sparse graphs.

For the models considered in this text, one way to achieve parsimony is through the removal of additional interaction terms with respect to those implied by the relevant graphical model. However, it is always desirable that models are interpretable, and thus an important issue concerns the interpretation of the additional constraints. From this perspective, when the interest is in independence relationships, it is somehow natural to investigate the identification of constraints admitting an interpretation of this kind. Relevant work in this area has focused on the implementation of *context-specific independencies*, i.e., conditional independencies that are not satisfied for every value of the conditioning variables but that only hold in specific contexts; see Boutilier *et al.* (1996), Corander (2003), Højsgaard (2004), Nyman *et al.* (2014), La Rocca and Roverato (2017) and references therein.

Context-specific independencies represent natural additional restrictions in the context of conditional independence models, but they do not suit well models of marginal independence. A natural approach for specifying parsimonious submodels of marginal independence models concerns the implementation of marginal independencies that hold between dichotomized versions of the variables. Indeed, one can notice that collapsing two or more levels of a non-binary variable into a single level can be regarded as a special type of marginalization. Accordingly, independence relationships that appear after some levels of a variable are collapsed into a single level can be regarded as special kinds of marginal independencies. The log-mean linear parametrization satisfies upward compatibility, but for variables with more than two levels it also satisfies a stronger property named *dichotomization invariance* (Roverato, 2015). This implies that marginal independencies between dichotomized versions of the variables correspond to zero log-mean linear terms. Interestingly, the bidirected graph of the model can be enlarged to accommodate such additional relationships, which can be read from the enlarged graph through the usual Markov properties. This approach is illustrated with an example, and we refer to Roverato (2015) for a formal presentation of this issue.

Table 5.3 gives a $4 \times 3 \times 3 \times 2$ contingency table originally described by Madsen (1976) that refers to a sample of 1681 residents of Copenhagen (see also Agresti, 2013, exercise 8.28). The variables are Y_1 = type of housing (levels: Ap = apartments, At = atrium houses, Te = terraced houses, To = tower blocks), Y_2 = feeling of influence on apartment management (levels: L_2 = low, M_2 = medium, H_2 = high), Y_3 = satisfaction with housing conditions (levels: L_3 = low, M_3 = medium, H_3 = high) and Y_4 = degree of contact with other residents (levels: L_4 = low, H_4 = high).

An exhaustive search within the family of bidirected graph models led to the selection of the graph in Figure 5.4 which encodes the structural relationship $Y_3 \perp\!\!\!\perp Y_4$ (Roverato, 2015). The selected model has deviance 5.13 on 2 degrees of freedom (*p*-value = 0.08). Parsimony is not achieved by this model that has 69 parameters, to be compared

Table 5.3 Data from Madsen (1976) on satisfaction with housing.

| Housing | Influence | Contact: Low | | | Contact: High | | |
| | | Satisfaction | | | Satisfaction | | |
		Low	Medium	High	Low	Medium	High
Tower blocks	Low	21	21	28	14	19	37
	Medium	34	22	36	17	23	40
	High	10	11	36	3	5	23
Apartments	Low	61	23	17	78	46	43
	Medium	43	35	40	48	45	86
	High	26	18	54	15	25	62
Atrium houses	Low	13	9	10	20	23	20
	Medium	8	8	12	10	22	24
	High	6	7	9	7	10	21
Terraced houses	Low	18	6	7	57	23	13
	Medium	15	13	13	31	21	13
	High	7	5	11	5	6	13

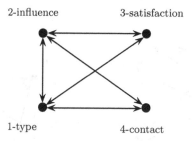

Figure 5.4 Housing data: bidirected graph of the selected model.

with the dimension of the saturated model that is equal to 71. Notice also that the analysis of these data in the context of undirected graph models would not give a more parsimonious model because no pairwise conditional independence can be identified on the basis of the asymptotic chi-squared distribution of the deviance.

Let Y_v be a variable with state space $\mathcal{I}_v = \{0, 1, \ldots, d_v\}$ and restricted state space $\mathcal{J}_v = \{1, \ldots, d_v\}$. Every level $i \in \mathcal{I}_v$ of Y_v is associated with a binary variable obtained by collapsing all the levels, with the exception of i, into a single level. More formally we let

$$X_v^i = \begin{cases} 1 & \textit{if } Y_v = i \\ 0 & \textit{otherwise.} \end{cases}$$

This gives $|\mathcal{I}_v| = d_v + 1$ binary variables, and one can replace Y_v with this collection of binary variables with no loss of information. On the other hand, this set of variables is redundant because any subset of d_v of such binary variables is sufficient to recover Y_v. Hence, we replace Y_v with the vector of binary variables indexed by its restricted state space; that is, we let $X_{\mathcal{J}_v} = (X_v^1, \ldots, X_v^{d_v})$ and call this vector a *binary expansion* of Y_v. The dichotomization invariance property of the log-mean linear parameterization implies that, if the restricted state space used to expand the variables is the same used in the computation of the log-mean linear parameter, then independence relationships between the entries of the binary expansions are equivalent to zero constraints in the log-mean linear parameters. Furthermore, it is possible to represent the independence structure through an *expanded bidirected graph* that is a bidirected graph where one or more variables are replaced by their binary expansions. This graph provides a complete description of the constraints characterizing the submodel, which are all feasible of an interpretation in terms of marginal independencies. Figure 5.5 shows

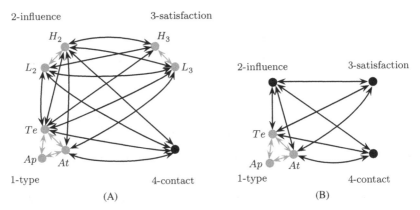

Figure 5.5 Housing data: expanded bidirected graphs.

two expanded bidirected graphs for the housing data where, to improve readability, the gray color is used to depict the complete subgraphs of the expanded variables.

The restricted state space is obtained by removing one level from $\mathcal{I}_v$, and the choice of such baseline level is arbitrary. Likewise, one can replace any variable Y_v with one binary vector arbitrarily chosen among $d_v + 1$ different, but equivalent, binary expansions. One drawback of this approach is therefore that the selected model depends on the chosen baseline level. On the other hand, sometimes there exists a natural way to choose the baseline level. For instance, for the housing data the variable Y_2 has three ordinal levels, $L_2 = $ low, $M_2 = $ medium, $H_2 = $ high; and the central level *medium* is a natural baseline because it makes sense to collapse it with either the level *low* or the level *high* leading to the binary expansions $(X_2^{L_2}, X_2^{H_2})$. Similarly, a natural binary expansion of Y_3 is $(X_3^{L_3}, X_3^{H_3})$. The variable Y_1 has no natural baseline and in the application presented below the level *tower blocks* is set as baseline thereby obtaining the binary expansion $(X_1^{Ap}, X_1^{At}, X_1^{Te})$. The variable Y_4 is binary and, therefore, it cannot be expanded.

Roverato (2015) applied this approach to the housing data and identified the model with graph in Figure 5.5A, which has deviance equal to 34.34 on 23 degrees of freedom (*p*-value = 0.06). This submodel is much more parsimonious than the model in Figure 5.4 with 21 additional vanishing log-mean linear terms that can be interpreted as $X_1^{Ap} \perp\!\!\!\perp (Y_2, Y_3, Y_4)$ and $X_1^{At} \perp\!\!\!\perp (X_2^{L_2}, X_3^{H_3})$. The graph in Figure 5.5A fully identifies the selected model, but has the disadvantage of being dense and thus difficult to read. For this reason, we also give the graph in

Figure 5.5B, where only variable Y_1 is expanded. The latter graph fails to represent the independence $X_1^{At} \perp\!\!\!\perp (X_2^{L_2}, X_3^{H_3})$, but may be regarded as a good compromise between readability of the graphical representation and completeness of the information provided by the graph.

A relevant feature of this approach is that the variables to be expanded do not need to be defined a priori, but they can be specified after a model has been selected from data. For instance, the set of variables to expand can be chosen so as to optimize the trade-off between readability of the graph and the need to explicitly represent the learnt independencies involving every single expanded variable, as shown in the application to the housing data. In this sense, it is possible to see the operation of variable expansion in the context of an interactive display, whereby the analyst could select the nodes to be expanded by, for instance, clicking on them.

6

Directed Acyclic and Regression Graph Models

In this chapter we introduce graphs containing directed edges, which are used to deal with asymmetric relationships between variables. We first consider statistical models associated with a family of graphs, called directed acyclic graphs, whose edge set only comprises directed edges. These models have a long history that may be traced back to the pioneering work of Wright (1921) on path analysis, and they are also commonly known in the literature as *Bayesian networks* (Pearl, 1986). Graphical models based on directed acyclic graphs constitute a very general framework. They can be used to represent several classical statistical formalisms such as linear structural equation models, models for factor analysis and various types of regression and latent variable models. Furthermore, they have proved useful in a wide range of applications including expert systems (Pearl, 1988; Cowell *et al.*, 1999), causal analysis (Spirtes *et al.*, 2000; Lauritzen, 2001; Pearl, 2009) and machine learning (Barber, 2012). Next, we will consider models for graphs, called regression graphs, that comprise the three kinds of edges: undirected, bidirected and directed. In a directed graph arrows point from vertices associated with explanatory variables to vertices associated with response variables. For example, the graph $a \to b \to c$ implies a relationship between variables where Y_a is a purely explanatory variable, Y_b is an intermediate variable, being both an explanatory variables of Y_c and a response of Y_a, and Y_c is a primary response. Regression graph models extend this idea to a multivariate regression setting (Cox and Wermuth, 1993, 1996; Wermuth and Sadeghi, 2012; Wermuth and Cox, 2015). In this approach, variables are arranged in a sequence of blocks ordered on the basis of subject matter considerations. Every block contains variables which are multiple responses of the previous blocks and explanatory of the following blocks. The first block has no incoming arrows, and therefore it contains purely explanatory variables, also called context variables. Likewise, the last block contains the primary responses. The variables within each block are considered to be on an equal standing, with the relationships between context variables represented by an undirected graph and those in each block of response variables by a bidirected graph. Graphical models for regression graphs form a flexible class that includes, as special cases, undirected, bidirected and directed acyclic graph models. Furthermore, they have close connections with other relevant families of graphical models such as models for multivariate regression chain graphs (Drton, 2009) and acyclic directed mixed graphs (Richardson, 2003).

In the parameterization of regression graph models we will follow the approach of La Rocca and Roverato (2017). Alternative approaches available in the literature are either based on a collection of marginal and conditional probabilities called (generalized) Möbius parameters (Drton, 2009; Evans and Richardson, 2010, 2014) or on the theory of marginal log-linear models (Bergsma and Rudas, 2002; Bartolucci et al., 2007; Evans and Richardson, 2013; Marchetti and Lupparelli, 2011; Rudas et al., 2010).

6.1 Directed Acyclic Graphs

Let $G = (V, E)$ be a graph. A directed edge is an ordered pair $(a, b) \in E$ representing an asymmetric relationship between the two vertices $a, b \in V$. The edge (a, b) is then depicted as an arrow pointing from a to b, $a \to b$, and in this case a is called a *parent* of b whereas b is a *child* of a. A graph with only directed edges is called *directed*, and when we want to highlight that an edge set only comprises directed edges we write it as $E^\to$. All the graphs we consider are acyclic. Specifically, a *directed acyclic graph* (DAG) is a directed graph with no directed cycles in the sense that it is not possible to return to any vertex by following the directions of the arrows; see Figure 6.1 for an example of a DAG.

A path is defined for directed graphs like in the undirected case, in the sense that no role is played by the directions of its arrows. Formally, in a directed graph a *path* with endpoints a and b is a sequence $a = v_0, \ldots, v_k = b$ of distinct vertices such that either $v_{r-1} \to v_r$ or $v_{r-1} \leftarrow v_r$ for all $r = 1, \ldots, k$. Examples of paths in the graph of Figure 6.1 are (b, e, g, f), (b, e, c, a, d, f) and (a, d, f, g, h). A non-endpoint vertex z of a path is called a *collider* on the path if the

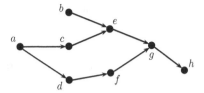

Figure 6.1 Example of directed acyclic graph (DAG).

edges of the path meet head-to-head at z, and it is a *non-collider* otherwise. Hence, the vertex e is a collider on the path (b,e,c,a,d,f), whereas c, a and d are non-collider vertices on the same path. Conversely, e is a non-collider vertex on the path (b,e,g,f). When all the edges of the path $a = v_0, \ldots, v_k = b$ are arrows pointing in the direction of b, that is $v_{r-1} \to v_r$ for all $r = 1, \ldots, k$, then the path is said to be directed from a to b. Thus, (a,d,f,g,h) is the only directed path among those considered in the example above.

We now introduce some relevant sets of vertices of a DAG $G = (V,E)$. Given a vertex $b \in V$, the set of all $v \in V$ such that $v \to b$ is the set of parents of b, denoted by $\text{pa}_G(b)$. If there is a directed path from a to b then the vertex b is called a *descendant* of a, and the set of descendants of a is denoted by $\text{de}_G(a)$. Finally, the set of *non-descendants* of a vertex a is defined as $\text{nd}_G(a) = V \backslash (\text{de}_G(a) \cup \{a\})$. For example, in the graph of Figure 6.1, $\text{pa}(g) = \{e,f\}$, $\text{de}(c) = \{e,g,h\}$, whereas $\text{nd}(c) = \{a,b,d,f\}$. Recall that we omit the subscript when it is clear from the context which graph is being considered.

The vertices of a DAG can always be *well-ordered* so that if two vertices form an edge, the arrow points from the lower to the higher of the two vertices with respect to the ordering. Notice that a DAG may not have a unique well-ordering. For example, two well-orderings of the vertices of the graph in Figure 6.1 are (a,b,c,d,e,f,g,h) and (b,a,d,f,c,e,g,h). If a graph G is well-ordered, the set of *predecessors* of a vertex a, denoted by $\text{pr}_G(a)$, is the set of those vertices that have a lower number than a. Hence, $\text{pr}(f) = \{a,b,c,d,e\}$ in the well-ordering (a,b,c,d,e,f,g,h), whereas in the well-ordering (b,a,d,f,c,e,g,h) we have $\text{pr}(f) = \{b,a,d\}$. It follows from the definition of descendant that for any vertex $a \in V$ of a well-ordered DAG it holds that $\text{pr}(a) \cap \text{de}(a) = \emptyset$ and therefore that $\text{pr}(a) \subseteq \text{nd}(a)$.

6.2 Markov Properties for Directed Acyclic Graphs

As for the two types of graphs considered so far, also for DAGs there exist a Markov property named pairwise that concerns independence relationships between pairs of variables, a local Markov property focusing on the relationship between every single variable and all the remaining variables, and a more general global Markov property involving independencies identified through a separation criterion. However, in this case an additional *ordered* Markov property is available. It applies to well-ordered DAGs, and suits the case when background knowledge on the causal or temporal ordering of the variables allows one to specify the directions of the arrows, and thus an ordering of the vertices.

The Directed Pairwise Markov Property

The probability distribution of a random vector Y_V is said to obey the *directed pairwise Markov property* with respect to the DAG $G = (V, E^{\rightarrow})$ if for every pair $a, b \in V$ of non-adjacent vertices with $b \in \mathrm{nd}(a)$ it holds that

$$Y_a \perp\!\!\!\perp Y_b | Y_{\mathrm{nd}(a) \setminus \{b\}}.$$

Examples of conditional independencies implied by the pairwise Markov property for the DAG of Figure 6.1 are $Y_h \perp\!\!\!\perp Y_a | Y_{\{b,c,d,e,f,g\}}$, $Y_g \perp\!\!\!\perp Y_c | Y_{\{a,b,d,e,f\}}$ and $Y_a \perp\!\!\!\perp Y_b$. Notice that this property is less interesting than the pairwise properties for undirected and bidirected graphs because the specification of the conditioning set is less natural.

The Directed Local Markov Property

The probability distribution of a random vector Y_V is said to obey the *directed local Markov property* with respect to the DAG $G = (V, E^{\rightarrow})$ if for every $a \in V$ it holds that

$$Y_a \perp\!\!\!\perp Y_{\mathrm{nd}(a) \setminus \mathrm{pa}(a)} | Y_{\mathrm{pa}(a)}.$$

Hence, for the DAG of Figure 6.1 the local Markov property implies, for instance, $Y_g \perp\!\!\!\perp Y_{\{a,b,c,d\}} | Y_{\{e,f\}}$, $Y_f \perp\!\!\!\perp Y_{\{a,b,c,e\}} | Y_d$ and $Y_a \perp\!\!\!\perp Y_b$.

The Ordered Markov Property

The probability distribution of a random vector Y_V is said to obey the *ordered Markov property* with respect to the well-ordered DAG $G = (V, E^\rightarrow)$ if for every $a \in V$ it holds that

$$Y_a \perp\!\!\!\perp Y_{\text{pr}(a) \setminus \text{pa}(a)} \mid Y_{\text{pa}(a)}.$$

For the DAG in Figure 6.1, the ordered Markov property with respect to the well-order (a, b, c, d, e, f, g, h) implies, for instance, $Y_e \perp\!\!\!\perp Y_{\{a,d\}} \mid Y_{\{b,c\}}$ and $Y_c \perp\!\!\!\perp Y_b \mid Y_a$.

The ordered Markov property replaces all non-descendants of a, $\text{nd}(a)$, in the local Markov property by its predecessors $\text{pr}(a)$ in some given well-ordering of the vertices. By noticing that $\text{pr}(a) \subseteq \text{nd}(a)$ for every $a \in V$ and for every well-ordering of the vertices, it follows immediately that the local directed Markov property implies the ordered Markov property with respect to any well-ordering of the vertices. In fact, we will see below that the ordered and the local Markov properties are equivalent because it is also true that the former implies the latter. This also shows that, if the distribution of Y is ordered Markov with respect to a given well-ordering, then it is ordered Markov with respect to all well-orderings of the vertices.

The Directed Global Markov Property

The global Markov property for DAGs requires the introduction of a kind of separation that is more involved than in the undirected and bidirected cases.

A path between a and b in a DAG $G = (V, E^\rightarrow)$ is *blocked* given a set $S \subseteq V$ if there is an inner vertex z of the path such that

(b1) either z is a non-collider on the path and $z \in S$,

(b2) or z is a collider on the path and neither z nor any descendant of z is in S; i.e., $(\text{de}(z) \cup \{z\}) \cap S = \varnothing$.

A path that is not blocked is a connecting path. More specifically, a path between a and b in G is *connecting* given a set $S \subseteq V$ if the following conditions are both satisfied:

(c1) every non-collider on the path is not contained in S;

(c2) every collider on the path is either in S or has a descendant in S.

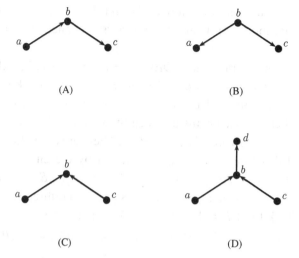

Figure 6.2 Four DAGs each containing a tripath (a,b,c).

Clearly, condition (b1) is satisfied for some vertex if and only if (c1) is not true and, likewise, condition (b2) holds true for some vertex if and only if (c2) is false.

We illustrate the notion of blocked and connecting paths by means of some examples. The most basic instance of a path containing inner vertices is called a *tripath* and has exactly three vertices, two of which are endpoints of the path and one is an inner vertex. Figure 6.2 displays four DAGs each containing a tripath (a, b, c). In the tripaths of Figure 6.2A and Figure 6.2B the inner vertex b is a non-collider and therefore these two paths are blocked given $S = \{b\}$, but they are both connecting paths given $S = \emptyset$. In the DAGs of Figure 6.2C and Figure 6.2D the inner vertex b is a collider and therefore in both DAGs the path (a, b, c) is blocked given $S = \emptyset$, but connecting given $S = \{b\}$. Furthermore, in the DAG of Figure 6.2D the vertex d is a descendant of the collider b and therefore the path (a, b, c) is connecting given $S = \{d\}$, and also given $S = \{b, d\}$. We now turn to the DAG of Figure 6.1. The path (b, e, g, f) is blocked, for instance, given any of the sets $S = \{e, g\}$ (the vertex e satisfies condition (b1)) $S = \emptyset$ and $S = \{a, c, d\}$ (in both cases the vertex g satisfies condition (b2)). On the other hand, the latter is a connecting path given, for instance, each of the sets $S = \{g\}, S = \{h\}$ and $S = \{a, c, d, h\}$. Finally, consider the path (b, e, c, a, d, f). Examples of sets blocking this path are $S = \emptyset$

(the vertex e satisfies condition (b2)) and $S = \{d, h\}$ (the vertex d satisfies condition (b1)). Conversely, this is a connecting path given, for instance, $S = \{e\}$, $S = \{h\}$ and $S = \{g, h\}$.

The notion of separation in a DAG $G = (V, E^{\rightarrow})$ was introduced by Pearl (1988). Two subsets $A, B \subseteq V$ are *d-separated* by $S \subseteq V$ if all paths from A to B are blocked by S. To see an example, in the DAG of Figure 6.1 there are two paths between the sets $A = \{b\}$ and $B = \{f\}$, specifically (b, e, g, f) and (b, e, c, a, d, f). These are both blocked given $S = \emptyset$ and therefore A and B are separated by the empty set. Other instances of separating sets are $S = \{e, d\}$, $S = \{a, d\}$ and $S = V \setminus \{b, f\} = \{a, c, d, e, g, h\}$. In the same DAG, three sets separating $A = \{b, e\}$ from $B = \{d, f\}$ are $\{a\}$, $\{c\}$ and $\{a, c\}$. Notice that neither $S = \emptyset$ nor $S = V \setminus (A \cup B) = \{a, c, g, h\}$ d-separate $\{b, e\}$ from $\{d, f\}$.

The probability distribution of a random vector Y_V is said to obey the *directed global Markov property* with respect to the DAG $G = (V, E^{\rightarrow})$ if for any triple of pairwise disjoint subsets $A, B, S \subseteq V$ such that S d-separates A from B in G it holds that

$$Y_A \perp\!\!\!\perp Y_B | Y_S.$$

Hence, if Y is globally Markov with respect to the DAG of Figure 6.1 then, for instance, we have $Y_b \perp\!\!\!\perp Y_f$, $Y_b \perp\!\!\!\perp Y_f | Y_{V \setminus \{b, f\}}$ and $Y_{\{b, e\}} \perp\!\!\!\perp Y_{\{d, f\}} | Y_{\{a, c\}}$.

The criterion of d-separation might be difficult to verify, and therefore the independence relationships implied by the global Markov property for DAGs are less intuitive than in the undirected and bidirected cases. However, this Markov property is important because Geiger and Pearl (1988) showed that it is complete, so that the d-separation criterion cannot be improved; see also Meek (1995).

The Factorization Property

Assume that the distribution of Y_V admits a density $f(\cdot)$ with respect to a product measure. Then the distribution of Y_V is said to *factorize* over the DAG $G = (V, E^{\rightarrow})$ if its density has the form

$$f(y) = \prod_{v \in V} f(y_v | y_{\mathrm{pa}(v)}), \tag{6.1}$$

where $f(\cdot | \cdot)$ are conditional densities.

In the directed case the relationships among alternative Markov properties, and between Markov properties and the factorization property, are simpler than in the undirected and bidirected cases. It can be shown (see, e.g. Cowell *et al.*, 1999, theorem 5.14) that if the distribution of Y admits a density, then the factorization property, the directed local, ordered and global Markov properties are all equivalent, i.e.,

$$(\text{factorization}) \iff (\text{global}) \iff (\text{local}) \iff (\text{ordered}).$$

In fact, Lauritzen *et al.* (1990, propositions 4 and 5) showed that the ordered, local and global Markov properties are equivalent even without assuming the existence of a density. The directed pairwise Markov property is implied by each of the four properties above, but the converse is not true in general. A sufficient condition for the equivalence of the pairwise Markov property with the other four properties is that the density of Y is positive (see Lauritzen, 1996, section 3.2.2).

Markov Equivalence

It is possible that two different graphs, under the respective Markov properties, encode the same collection of independence relationships. In this case the two graphs define the same statistical model and are said to be *Markov equivalent*. For example, the collection of independence relationships implied by the DAG in Figure 6.2A comprises the unique independence $Y_a \perp\!\!\!\perp Y_c | Y_b$. Also the DAG in Figure 6.2B only implies the relationship $Y_a \perp\!\!\!\perp Y_c | Y_b$ and therefore the two DAGs are Markov-equivalent. Furthermore, these two DAGs are both Markov-equivalent to the undirected graph $a — b — c$. The DAG in Figure 6.2A encodes the independence $Y_a \perp\!\!\!\perp Y_c$ and it is Markov-equivalent to the bidirected graph $a \leftrightarrow b \leftrightarrow c$. Markov equivalence is an important question in graphical modelling with relevant implications in many domains such as structure learning (Drton and Maathuis, 2017) and causal discovery (Spirtes *et al.*, 2000; Pearl, 2009). We do not provide additional details on this subject, and the interested reader is referred to Frydenberg (1990), Andersson *et al.* (1997), Pearl and Verma (1990), Volf and Studený (1999), Chickering (2002), Roverato (2005), Roverato and Studený (2006), Roverato and La Rocca (2006), Drton and Richardson (2008b), Ali *et al.* (2009), Wermuth and Sadeghi (2012) and references therein for a comprehensive presentation of the theory of Markov equivalence and related problems.

6.3 Regression Graphs

Up to now we have considered graphs consisting of a single type of
edge. Here we introduce mixed graphs whose edge set may contain
undirected, bidirected and directed edges. Our main focus is on the class
of *regression graphs* which belong to the wider family of *chain graphs*.
In a chain graph the vertex set can be partitioned into an ordered
sequence of pairwise disjoint *blocks*. Directed edges are not allowed
within blocks, and all the edges joining vertices belonging to different
blocks are arrows pointing from the lower to the higher of the two
blocks with respect to the ordering. The connected components of the
blocks are called *chain components*, and we denote the set of chain
components of a chain graph by $K = \{K_1, \ldots, K_s\}$. Notice that the
partition of the vertex set into blocks may not be unique, but the set of
chain components is unique and it does not depend on the block partition. A chain graph where all chain components are singletons is
a DAG.

A regression graph is a chain graph $G = (V, E)$ whose vertex set can
be partitioned into an ordered sequence of blocks $(T_0, T_1, \ldots, T_k)$ such
that the subgraph G_{T_0} induced by the first block T_0 is undirected,
whereas the subgraphs induced by the remaining blocks, G_{T_r} for
$r = 1, \ldots, k$, are all bidirected. The graph in Figure 6.3 is a regression
graph with blocks $T_0 = \{a, b, c, d, e\}$, $T_1 = \{f, g, h\}$ and
$T_2 = \{l, m, q, r\}$, whereas the set of chain components is

$$K = \{\{a, b, c\}, \{d, e\}, \{f, g, h\}, \{l, m, q, r\}\}.$$

The notion of parent, given for DAGs, can also be applied to chain
graphs. If $G = (V, E)$ is a chain graph then the parents of the vertex

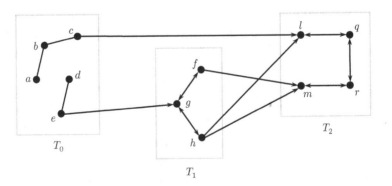

Figure 6.3 Example of regression graph with blocks (T_0, T_1, T_2).

$a \in V$, $\mathrm{pa}_G(a)$, is the set of vertices $v \in V$ such that G contains the edge $v \to a$. Furthermore, the set of parents of a subset $A \subseteq V$ that is defined as $\mathrm{pa}_G(A) = \cup_{a \in A}\mathrm{pa}(a) \backslash A$. Hence, in the graph of Figure 6.3 it holds, for example, that $\mathrm{pa}(l) = \{c, h\}$, $\mathrm{pa}(q) = \varnothing$ and $\mathrm{pa}(\{l, q\}) = \{c, h\}$. Furthermore, $\mathrm{pa}(T_2) = \{c, f, h\}$, $\mathrm{pa}(T_1) = \{e\}$ and $\mathrm{pa}(T_0) = \varnothing$. For a regression graph with block ordering $(T_0, T_1, \ldots, T_k)$, we define the sets of *predecessors* of a bidirected block T_r as $\mathrm{pr}_G(T_r) = T_0 \cup \ldots \cup T_{r-1}$ for $r = 1, \ldots, k$. Thus, in the graph of Figure 6.3 we have $\mathrm{pr}(T_2) = \{a, b, c, d, e, f, g, h\}$ and $\mathrm{pr}(T_1) = \{a, b, c, d, e\}$.

6.4 Markov Properties for Regression Graphs

The Markov property for regression graphs considered here unifies the global Markov properties for undirected graphs, bidirected graphs and DAGs. Consider a random vector Y_V and a regression graph $G = (V, E)$ with ordered sequence of blocks $(T_0, T_1, \ldots, T_k)$. The distribution of Y_V obeys the *ordered regression graph Markov property* with respect to $G = (V, E)$ when

(RG.d) $Y_A \perp\!\!\!\perp Y_{\mathrm{pr}(T_r) \backslash \mathrm{pa}(A)} | Y_{\mathrm{pa}(A)}$ for all $A \subseteq T_r$ and every $r = 1, \ldots, k$;

(RG.b) the distribution of $Y_{T_r} | Y_{\mathrm{pr}(T_r)}$ obeys the dual global Markov property with respect to the induced subgraph G_{T_r}, for every $r = 1, \ldots, k$;

(RG.u) the distribution of Y_{T_0} obeys the global Markov property with respect to the induced subgraph G_{T_0}.

Conditions (RG.b) and (RG.u) concern the independence relationships implied by missing bidirected and undirected edges, respectively. Condition (RG.d) is well-defined when both $A \neq \varnothing$ and $\mathrm{pr}(T_r) \backslash \mathrm{pa}(A) \neq \varnothing$, and it gives the independence relationships implied by missing arrows. When all blocks are singletons, (RG.d) simplifies to the ordered directed Markov property for DAGs.

This formulation of the Markov property for regression graphs was first given in Marchetti and Lupparelli (2011), where it is shown that models defined by the above conditions do not depend on the chosen block ordering of variables; see also Drton (2009). The ordered Markov property is the only Markov property that we introduce for regression graphs, and in the following we will shortly say that a distribution is Markov with respect to a regression graph. The reader interested in the alternative Markov properties for regression graphs can find the theory relative to the pairwise Markov property in Sadeghi and Wermuth (2016). A global Markov property for regression graphs can be obtained by exploiting the Markov properties defined for other classes of models for mixed graphs, which have close connections with regression graphs. Several classes of graphical models for mixed graphs are available; see Sadeghi and Lauritzen (2014) for an overview. A relevant subfamily of models for mixed graphs is known in the literature as *multivariate regression chain graphs* (MRCGs), but it was also called *chain graphs of type IV* by Drton (2009). MRCGs belong to the wider class of *acyclic directed mixed graphs* (ADMGs), and there exists a global Markov property for ADMG models. This is based on the notion of *m*-separation, that is an extension of the *d*-separation criterion (Richardson and Spirtes, 2002; Richardson, 2003; Sadeghi and Lauritzen, 2014). An MRCG is a regression graph without the undirected block, i.e., with $T_0 = \varnothing$, and it can be shown that in this case the ordered Markov property for regression graphs is equivalent to the global Markov property for MRCGs (Drton, 2009; Marchetti and Lupparelli, 2011). This equivalence can be extended to the case where $T_0 \neq \varnothing$ by considering the global Markov property for the undirected component and the ADMG Markov property for the distribution of Y_V conditional on the undirected component.

Assume that the distribution of Y_V admits a density $f(\cdot)$ with respect to a product measure. If Y is Markov with respect to a regression graph $G = (V, E)$ then the density factorizes as,

$$f(y) = \prod_{K \in \mathcal{K}} f(y_K | y_{\text{pa}(K)}); \qquad (6.2)$$

see Lauritzen (1996, section 3.2.3) and Drton (2009). This factorization is essentially identical to the factorization (6.1) for DAGs. However, in general the factorization (6.2) is not equivalent to the Markov property for regression graphs because it typically implies only a subset of the independence relationships implied by the Markov property.

6.5 On the Interpretation of Models defined by Regression Graphs

The family of chain graphs first introduced in the graphical model literature has two types of edges, directed and undirected (Lauritzen and Wermuth, 1989). Accordingly, the classical Markov property for chain graphs is a natural generalization of both the Markov property for undirected graphs and DAGs (Frydenberg, 1990). One motivation for extending DAGs to graphs including undirected edges was to allow the modelling of "symmetric associations" and "simultaneous responses" (Lauritzen and Richardson, 2002). More recently, chain graph models have been extended to include a third type of edge, i.e., bidirected arrows, to express a second type of symmetric relationship between variables, and today four different Markov properties are available for chain graphs (Frydenberg, 1990; Wermuth and Lauritzen, 1990; Andersson *et al.*, 2001; Drton, 2009).

We focus on regression graphs because, whenever it is reasonable to assume the existence of a partial ordering of the variables, these models play a central role among the existing families of chain graph models; see Wermuth and Sadeghi (2012) and Wermuth and Cox (2015) for further discussion on this issue. As far as interpretation is of concern, in regression graph models the distribution of every block conditioned on the previous blocks is a multivariate regression and, as such, it focuses on the marginal distributions of response variables (see, e.g. Whittaker, 1990, section 10.5). To fix ideas, consider the model implied by the graph in Figure 6.3. In the multivariate regression with response Y_{T_2} one can restrict the attention to specific subsets of responses, for example Y_l and $Y_{\{m,q\}}$, and apply (RG.d) to read from the graph that $Y_l \perp\!\!\!\perp Y_{\{a,b,e,d,f,g\}} | Y_{\{c,h\}}$ and that $Y_{\{m,q\}} \perp\!\!\!\perp Y_{T_0 \cup \{g\}} | Y_{\{f,h\}}$. Likewise, in the regression with response Y_{T_1} it holds for example that $Y_{\{g,h\}} \perp\!\!\!\perp Y_{\{a,b,c,d\}} | Y_e$.

Regression graph models also model the possible presence of residual associations among the responses. This is done through a bidirected graph and, in this regard, it is worth mentioning that bidirected edges are feasible of an interpretation in terms of latent variables; see, e.g., Richardson and Spirtes (2002) and Evans (2016). For example, the DAG model with graph

$$a \leftarrow u_1 \rightarrow b \leftarrow u_2 \rightarrow c$$

implies the marginal independence of Y_a and Y_c so that, if the variables for the vertices u_1 and u_2 are unobserved, then the independence structure of the observed variables is provided by the bidirected graph

$$a \leftrightarrow b \leftrightarrow c.$$

Another example is given in the application of Section 5.5 where the DAG in Figure 5.3C implies, for the four observed variables, the independence structure encoded by the regression graph

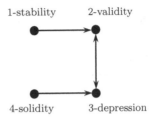

that is also Markov equivalent to the bidirected graph of Figure 5.3B.

We now turn to the first block T_0 of context variables. For background variables, which are purely explanatory, no incoming arrow is assumed, and thus the association structure may be suitably represented by means of an undirected graph that provides insight into the conditional independence relationship holding among these variables.

6.6 The Log-hybrid Linear Parameterization

As a first step towards the parameterization of regression graph models, we consider the case where V is partitioned into two subsets, which we also call blocks, T and U. We write this partition as an ordered pair (U, T) to underline the different role played by U, that indexes a vector of explanatory variables Y_U, with respect to T, that indexes a vector of response variables Y_T. Relative to this partition, we describe a parameterization that unifies the log-linear and the log-mean linear approaches, in the sense that independence relationships among explanatory variables correspond to the vanishing of interactions associated with cross-product ratios of the "conditional" type, $\mathrm{cpr}_{cc}(\cdot)$, whereas independence relationships among response variables correspond to the vanishing of interactions associated with cross-product ratios of the "marginal" type, $\mathrm{cpr}_{mm}(\cdot)$. The novel feature of this parameterization concerns independence relationships between variables belonging to different blocks, which correspond to the vanishing of interactions associated with cross-product ratios of the "marginal-conditional" type, $\mathrm{cpr}_{mc}(\cdot)$.

In the following, we first introduce a generalization of the zeta matrix, and then we show how this can be used in the computation of the parameters of interest and in the derivation of their properties.

As usually done in this text, the binary case is considered in detail whereas for the general case of polytomous variables only a list of technical results is provided.

The Matrices $\mathbb{I}_U \otimes \mathbb{Z}_T$ and $\mathbb{I}_U \otimes \mathbb{M}_T$

In order to facilitate the computation of some relevant quantities, it is useful to extend the zeta and the Möbius matrices so as to explicitly take into account the asymmetric block partitioning of the vertices. This is done by associating an identity matrix with the block U. Initially, we consider singleton sets, and denote by $\mathbb{I}_v$ the identity matrix associated with $v \in V$, that is

$$\begin{array}{cc} & \begin{array}{cc} \varnothing & \{v\} \end{array} \\ \begin{array}{c} \varnothing \\ \{v\} \end{array} & \begin{pmatrix} 1 & 0 \\ 0 & 1 \end{pmatrix} \end{array} = \mathbb{I}_v,$$

where the matrix is displayed with its row and column indexes. Next, we define the identity matrix associated with a subset $A \subseteq V$ as

$$\mathbb{I}_A = \otimes_{v \in A} \mathbb{I}_v,$$

with the convention that $\mathbb{I}_\varnothing = 1$. The rows and columns of $\mathbb{I}_A$ are indexed by the subsets of A, so that the entry of $\mathbb{I}_A$ indexed by the pair $D, E \subseteq A$ is $(\mathbb{I}_A)_{D,E} = 1(D = E)$; it is also a well-known fact that $\mathbb{I}_A^{-1} = \mathbb{I}_A$. The set U can thus be associated with $\mathbb{I}_U$, the set T with $\mathbb{Z}_T$, and the pair (U, T) with the Kronecker product of $\mathbb{I}_U$ and $\mathbb{Z}_T$; i.e., $\mathbb{I}_U \otimes \mathbb{Z}_T$. By applying the properties of the Kronecker product, one easily obtains that $(\mathbb{I}_U \otimes \mathbb{Z}_T)^{-1} = \mathbb{I}_U \otimes \mathbb{M}_T$ and, furthermore, Proposition 3.1 can be extended to give an explicit formulation of the entries of $\mathbb{I}_U \otimes \mathbb{Z}_T$ and $\mathbb{I}_U \otimes \mathbb{M}_T$.

Proposition 6.1 *Given a partition (U, T) of V, let $\mathbb{I}_U$ be the identity matrix associated with U, and let $\mathbb{Z}_T$ and $\mathbb{M}_T$ be the zeta and the Möbius matrix, respectively, associated with T. Then, for every pair of subsets $D, E \subseteq V$*

$$(\mathbb{I}_U \otimes \mathbb{Z}_T)_{D,E} = 1(U \cap E \subseteq D \subseteq E),$$

whereas

$$(\mathbb{I}_U \otimes \mathbb{M}_T)_{D,E} = (-1)^{|E \setminus D|} \, 1(U \cap E \subseteq D \subseteq E). \qquad \square$$

Proof. We first show that the pair of subsets $D, E \subseteq V$ satisfies the condition (i) $(U \cap D = U \cap E) \wedge (T \cap D \subseteq T \cap E)$ if and only if it satisfies (ii) $(U \cap E \subseteq D \subseteq E)$. Condition (i) implies (ii) because the identity $U \cap E = U \cap D$ implies $U \cap E \subseteq D$, whereas $U \cap D = U \cap E$ together with $T \cap D \subseteq T \cap E$ implies $(U \cap D) \cup (T \cap D) \subseteq (U \cap E) \cup (T \cap E)$ and therefore that $D \subseteq E$. To show the reverse implication, i.e., that (ii) implies (i), we notice that $D \subseteq E$ implies $T \cap D \subseteq T \cap E$ whereas $D \subseteq E$ implies $U \cap D \subseteq U \cap E$ and $U \cap E \subseteq D$ implies $U \cap E \subseteq U \cap D$ so that $U \cap D = U \cap E$.

This result can be used to compute the entries of $\mathbb{I}_U \otimes \mathbb{Z}_T$ because

$$
\begin{aligned}
(\mathbb{I}_U \otimes \mathbb{Z}_T)_{D,E} &= (\mathbb{I}_U)_{U \cap D, U \cap E} \times (\mathbb{Z}_T)_{T \cap D, T \cap E} \\
&= 1(U \cap D = U \cap E) \times 1(T \cap D \subseteq T \cap E) \\
&= 1(U \cap E \subseteq D \subseteq E)
\end{aligned}
$$

where the entries of $\mathbb{Z}_T$ are given in Proposition 3.1. Similarly,

$$
\begin{aligned}
(\mathbb{I}_U \otimes \mathbb{M}_T)_{D,E} &= 1(U \cap D = U \cap E) \times (-1)^{|(T \cap E) \setminus (T \cap D)|} 1(T \cap D \subseteq T \cap E) \\
&= (-1)^{|E \setminus D|} 1(U \cap E \subseteq D \subseteq E),
\end{aligned}
$$

where $(T \cap E) \setminus (T \cap D) = E \setminus D$ because, as shown above, when $U \cap E \subseteq D \subseteq E$ it holds that $U \cap D = U \cap E$. $\qquad\square$

The Binary Case

Consider a vector Y of binary random variables with probability parameter π. We have seen that undirected graph models are suitably parameterized by the log-linear expansion $\lambda = \mathbb{M}^T \log \pi$ of π, whereas for bidirected graph models it is more convenient to employ the log-linear expansion $\gamma = \mathbb{M}^T \log \mu$ of the mean parameter $\mu = \mathbb{Z}\pi$. Every entry $\pi_D = p(Y_D = 1, Y_{V \setminus D} = 0)$ of π is a probability involving all the variables in Y_V, whereas every entry $\mu_D = p(Y_D = 1)$ of μ is a probability from the marginal distribution of Y_D. We now introduce a further parameterization that combines the features of π and μ. This is associated with the partition (U, T), and consists of probabilities involving all the variables in Y_U but only certain subvectors of Y_T. Formally, the *hybrid probability-mean parameter* of Y_V, or *hybrid parameter* for short, with respect to the partition (U, T) is the vector $\pi^{(U,T)} = (\pi_D^{(U,T)})_{D \subseteq V}$ with entries

$$\pi_D^{(U,T)} = p(Y_{U \cap D} = 1, Y_{U \setminus D} = 0, Y_{T \cap D} = 1)$$
$$= p(Y_D = 1, Y_{U \setminus D} = 0). \tag{6.3}$$

The hybrid parameter $\pi^{(U,T)}$ includes, as special cases, both the probability and the mean parameters; specifically, $\pi = \pi^{(V,\varnothing)}$ and $\mu = \pi^{(\varnothing,V)}$. Furthermore, $\pi^{(U,T)}$ can be computed from π as $\pi^{(U,T)} = (\mathbb{I}_U \otimes \mathbb{Z}_T)\,\pi$, that for $U = \varnothing$ simplifies to $\mu = \mathbb{Z}_V \pi$.

Proposition 6.2 *Let Y_V be a binary random vector with probability parameter π. The hybrid parameter $\pi^{(U,T)}$ of Y_V, with respect to the partition (U, T) of V, can be computed as*

$$\pi^{(U,T)} = (\mathbb{I}_U \otimes \mathbb{Z}_T)\,\pi \text{ so that } \pi = (\mathbb{I}_U \otimes \mathbb{M}_T)\,\pi^{(U,T)}.$$

Proof. We have to show that for every $D \subseteq V$ it holds that

$$\pi_D^{(U,T)} = \sum_{E \subseteq V} (\mathbb{I}_U \otimes \mathbb{Z}_T)_{D,E}\,\pi_E,$$

and the result follows from Proposition 6.1 because

$$\sum_{E \subseteq V} (\mathbb{I}_U \otimes \mathbb{Z}_T)_{D,E} \pi_E = \sum_{E \subseteq V} \mathbb{1}(U \cap E \subseteq D \subseteq E)\pi_E$$

$$= \sum_{F \subseteq T \setminus D} \pi_{D \cup F}$$

$$= \sum_{F \subseteq T \setminus D} p(Y_D = 1, Y_F = 1, Y_{(T \setminus D) \setminus F} = 0, Y_{U \setminus D} = 0)$$

$$= p(Y_D = 1, Y_{U \setminus D} = 0)$$

$$= \pi_D^{(U,T)}. \qquad \square$$

The parameters λ and γ are obtained from the log-linear expansion of π and μ, respectively, and it is therefore natural to consider the log-linear expansion of $\pi^{(U,T)}$ so as to obtain the *log-hybrid linear parameter* of Y (La Rocca and Roverato, 2017). This is denoted by $\varphi^{(U,T)}$, or simply by φ when it is clear which partition of V is being used, and is defined as

$$\varphi^{(U,T)} = \mathbb{M}^T \log \pi^{(U,T)}. \tag{6.4}$$

It follows from Proposition 6.2 that $\varphi = \mathbb{M}^T \log\{(\mathbb{I}_U \times \mathbb{Z}_T)\pi\}$. This shows that $\lambda = \varphi^{(V,\varnothing)}$ whereas $\gamma = \varphi^{(\varnothing,V)}$ and, furthermore, that π can be obtained from φ as $\pi = (\mathbb{I}_U \times \mathbb{M}_T)\exp(\mathbb{Z}^T \varphi)$ so that there is a one-to-one correspondence between π and φ. More precisely, the latter is a smooth bijection, and this implies that φ is a valid parameterization of the distribution of Y_V. The individual entries of φ can be computed as

$$\varphi_D = \sum_{D' \subseteq D} (-1)^{|D \setminus D'|} \log \pi_{D'}^{(U,T)} \quad \text{for every } D \subseteq V, \tag{6.5}$$

and, conversely,

$$\log \pi_D^{(U,T)} = \sum_{D' \subseteq D} \varphi_{D'} \quad \text{for every } D \subseteq V. \tag{6.6}$$

The importance of the log-hybrid linear parameterization stems from the fact that any independence relationship implied by the regression graph Markov property is equivalent to the vanishing of certain φ-terms. To illustrate the connection between zero entries of φ and independence relationships, we distinguish among three different cases. Firstly, we consider independence relationships between variables belonging to different blocks, and the rules to deal with this situation are stated in Theorem 6.3. Secondly, we focus on independencies holding among response variable and provide the relevant rules in Theorem 6.5. Finally, we consider independencies holding among context variables with the rules pertaining to this case given in Theorem 6.7.

In the Markov property for regression graphs, the independencies involving variables belonging to different blocks are those implied by missing arrows. The simplest case of this situation is when $|U| = 1$ and $|D| = 2$, say $U = \{b\}$ and $D = \{a, b\}$ with $a \in T$, so that $\pi_D^{(U,T)} = p(Y_a = 1, Y_b = 1)$ and (6.6) becomes

$$\log p(Y_a = 1, Y_b = 1) = \varphi_\emptyset + \varphi_a + \varphi_b + \varphi_{\{a,b\}}.$$

It is easy to see from (6.5) that $\varphi_\emptyset = \log p(Y_b = 0)$, whereas

$$\varphi_a = \log p(Y_a = 1 | Y_b = 0) \quad \text{and} \quad \varphi_b = \log \frac{p(Y_b = 1)}{p(Y_b = 0)}.$$

However, the main interest if for the two-way interaction $\varphi_{\{a,b\}}$ that coincides with the logarithm of the cross product ratio in (2.19),

$$\varphi_{\{a,b\}} = \log \frac{p(Y_a = 1, Y_b = 1)p(Y_b = 0)}{p(Y_a = 1, Y_b = 0)p(Y_b = 1)} = \log \text{cpr}_{mc}(a, a : b, b),$$

and it follows from Lemma 2.2 that $Y_a \perp\!\!\!\perp Y_b$ if and only if $\varphi_{\{a,b\}} = 0$. It is also worthwhile noticing that the interaction $\varphi_{\{a,b\}}$ can also be interpreted as the log-relative risk of the event $\{Y_a = 1\}$ with respect to the two groups identified by $\{Y_b = 1\}$ and $\{Y_b = 0\}$,

$$\varphi_{\{a,b\}} = \log \frac{p(Y_a = 1 | Y_b = 1)}{p(Y_a = 1 | Y_b = 0)}.$$

The connection between φ-terms and relative risks is discussed, with more generality, in Lupparelli and Roverato (2017).

Assume now that Y_U contains an arbitrary number of explanatory variables. Condition (RG.d) concerns independence relationships between variables Y_A, with $A \subseteq T$, and variables Y_B, with $B \subseteq U$, given the remaining variables in Y_U; that is $Y_A \perp\!\!\!\perp Y_B | Y_{U \setminus B}$. Consider, for example, the case where $A = \{a\}$ and $B = \{b\}$. The interaction $\varphi_{\{a,b\}}$ is equal to the logarithm of a conditional cross-product ratio,

$$\varphi_{\{a,b\}} = \log \frac{p(Y_a = 1, Y_b = 1, Y_{U \setminus \{b\}} = 0)p(Y_b = 0, Y_{U \setminus \{b\}} = 0)}{p(Y_a = 1, Y_b = 0, Y_{U \setminus \{b\}} = 0)p(Y_b = 1, Y_{U \setminus \{b\}} = 0)}$$

$$= \log \frac{p(Y_a = 1, Y_b = 1 | Y_{U \setminus \{b\}} = 0)p(Y_b = 0 | Y_{U \setminus \{b\}} = 0)}{p(Y_a = 1, Y_b = 0 | Y_{U \setminus \{b\}} = 0)p(Y_b = 1 | Y_{U \setminus \{b\}} = 0)}$$

$$= \log \mathrm{cpr}_{mc}(a, a : b, b | \emptyset, U \setminus \{b\}).$$

Hence, by Corollary 2.4, the relationship $Y_a \perp\!\!\!\perp Y_b | Y_{U \setminus \{b\}}$ implies that $\varphi_{\{a,b\}} = 0$. However, the vanishing of $\varphi_{\{a,b\}}$ is not sufficient for the reverse implication to hold true. Indeed, the independence $Y_a \perp\!\!\!\perp Y_b | Y_{U \setminus \{b\}}$ is equivalent to the vanishing of the interactions $\varphi_{\{a,b\} \cup C}$ for all $C \subseteq U \setminus \{b\}$. This is formally stated in the theorem given below, which generally deals with independence relationships between response variables and explanatory variables given the remaining explanatory variables. This result is an immediate consequence of the properties of cross product ratios and Möbius inversion and, in the context of regression graphs, it plays the same role as Theorem 4.2 in the context of undirected graphs and Theorem 5.2 in the context of bidirected graphs; see also La Rocca and Roverato (2017).

Theorem 6.3 *For a vector Y_V of binary variables with probability parameter $\pi > 0$, let $\varphi = \mathrm{M}^T \log \pi^{(U,T)}$, where $\pi^{(U,T)} = (\mathbb{I}_U \otimes \mathbb{Z}_T) \pi$ and $V = U \cup T$ is a partition of V. Then, for a pair of* *nonempty subsets $A \subseteq T$ and $B \subseteq U$ the following conditions are equivalent:*

(i) $Y_A \perp\!\!\!\perp Y_B | Y_{U \setminus B}$;

(ii) *for every $D \subseteq A \cup U$ such that both $D \cap A \neq \varnothing$ and $D \cap B \neq \varnothing$ it holds that*

$$\varphi_D = 0;$$

(iii) *for every $A' \subseteq A$, $B' \subseteq B$, and $C' \subseteq U \setminus B$, such that $A', B' \neq \varnothing$ it holds that*

$$\log \pi_{A' \cup B' \cup C'}^{(U,T)} - \log \pi_{B' \cup C'}^{(U,T)} - \log \pi_{A' \cup C'}^{(U,T)} + \log \pi_{C'}^{(U,T)} = 0.$$

$$(6.7)$$

Proof. The equivalence of (i) and (iii) follows from Corollary 2.4 because the identity (6.7) is equal to $\log \mathrm{cpr}_{mc}(A', A : B', B | C', C) = 0$ with $C = U \setminus B$. The equivalence of (ii) and (iii) is that stated in Lemma 3.3 when $\theta = \varphi$ and $\omega = \log \pi^{(U,T)}$. $\qquad\square$

Assume that the distribution of Y is Markov with respect to the graph in Figure 6.4. Hence, (RG.d) implies, for example, that $Y_e \perp\!\!\!\perp Y_{\{a,b\}} | Y_{\{c,d\}}$ and by Theorem 6.3 this relationship is satisfied if and only if $\varphi_D = 0$ for all $D \subseteq \{a,b,c,d,e\}$ such that both $e \in D$ and $D \cap \{a,b\} \neq \varnothing$. The same graph also implies $Y_{\{g,h\}} \perp\!\!\!\perp Y_{\{a,b,c,d\}}$ and this independence holds if and only if $\varphi_D = 0$ for all $D \subseteq \{a,b,c,d,g,h\}$ such that both $D \cap \{a,b,c,d\} \neq \varnothing$ and $D \cap \{g,h\} \neq \varnothing$.

We now turn to the analysis of the independence relationships between two subsets of response variables, which are considered conditionally on the explanatory variables. Hence, for a subset $U' \subseteq U$ we consider the event $\{Y_{U'} = 1, Y_{U \setminus U'} = 0\}$ and focus on the conditional

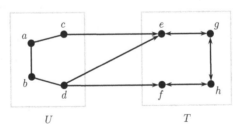

Figure 6.4 Regression graph with an undirected block $U = \{a, b, c, d\}$ and a bidirected block $T = \{e, f, g, h\}$.

distribution of $Y_T|\{Y_{U'} = 1, Y_{U\setminus U'} = 0\}$. Then, we denote the probability parameter of this distribution by $\pi^{T|U'} = (\pi_D^{T|U'})_{D\subseteq T}$ where, for $D \subseteq T$,

$$\pi_D^{T|U'} = p(Y_D = 1, Y_{T\setminus D} = 0|Y_{U'} = 1, Y_{U\setminus U'} = 0).$$

In the Markov property for regression graphs, independencies among response variables are implied by missing bidirected arrows, and therefore we compute the log-mean linear parameters

$$\gamma^{T|U'} = \mathbb{M}_T^T \log(\mathbb{Z}_T \, \pi^{T|U'}) \quad \text{for every} \quad U' \subseteq U. \tag{6.8}$$

The parameter $\gamma^{T|U'}$ can be used to establish "marginal" independence relationships in the distribution of $Y_T|\{Y_{U'} = 1, Y_{U\setminus U'} = 0\}$. More precisely, Theorem 5.2 implies that,

- for a pair of disjoint nonempty subsets A and B of T it holds that

$$Y_A \perp\!\!\!\perp Y_B|\{Y_{U'} = 1, Y_{U\setminus U'} = 0\}$$

- if and only if $\gamma_D^{T|U'} = 0$ for every $D \subseteq A \cup B$ such that $D \cap A \neq \emptyset$ and $D \cap B \neq \emptyset$.

Furthermore, $Y_A \perp\!\!\!\perp Y_B|Y_U$ if and only if $Y_A \perp\!\!\!\perp Y_B|\{Y_{U'} = 1, Y_{U\setminus U'} = 0\}$ for every $U' \subseteq U$, and this leads to the following equivalence:

- for a pair of disjoint nonempty subsets A and B of T it holds that

$$Y_A \perp\!\!\!\perp Y_B|Y_U$$

- if and only if $\gamma_D^{T|U'} = 0$ for every $D \subseteq A \cup B$ such that $D \cap A \neq \emptyset$ and $D \cap B \neq \emptyset$ and for every $U' \subseteq U$.

A relevant fact concerning the log-hybrid linear parameter φ is that it is related with the log-mean linear parameters in (6.8), through Möbius inversion, in such a way that it can be equivalent used to establish the independence relationships implied by condition (RG.b). The relevant Möbius inversion formula follows from the application of Lemma 3.4.

Proposition 6.4 *In the setting of Theorem 6.3, for $U' \subseteq U$ let $\gamma^{T|U'} = (\gamma_D^{T|U'})_{D\subseteq T}$ be the log-mean linear parameter of the distribution of $Y_T|\{Y_U = 1, Y_{U\setminus U'} = 0\}$. Then, for every nonempty subset $D \subseteq T$ it holds that*

$$\gamma_D^{T|U'} = \sum_{U'' \subseteq U'} \varphi_{D \cup U''}. \tag{6.9}$$

Proof. For every $U' \subseteq U$ and nonempty $D \subseteq V$ we can apply the definition of log-mean linear parameter in (5.2) to compute

$$\gamma_D^{T|U'} = \sum_{D' \subseteq D} (-1)^{|D \setminus D'|} \log p(Y_{D'} = 1 | Y_{U'} = 1, Y_{U \setminus U'} = 0) \tag{6.10}$$

$$= \sum_{D' \subseteq D} (-1)^{|D \setminus D'|} \log p(Y_{D'} = 1, Y_{U'} = 1, Y_{U \setminus U'} = 0) \tag{6.11}$$

$$= \sum_{D' \subseteq D} (-1)^{|D \setminus D'|} \log \pi_{U' \cup D'}^{(U,T)} \tag{6.12}$$

$$= \varphi_{[U']D}^{(U,T)}. \tag{6.13}$$

Because we have assumed $D \neq \varnothing$, then the sum in (6.10) has the same number of positive and negative terms. As a consequence, the sum does not change if, before taking the logarithm, every probability is multiplied by a positive constant, and we go from (6.10) to (6.11) by choosing such constant equal to $p(Y_{U'} = 1, Y_{U \setminus U'} = 0)$. Equation (6.12) is obtained by applying the definition of $\pi_{U' \cup D'}^{(U,T)}$ in (6.3) and, finally, (6.13) follows from the definition of $\varphi_{[U']D}^{(U,T)}$ in (3.7). Hence, the required result can be obtained by applying Lemma 3.4 to (6.13). □

The result of Proposition 6.4 can be applied to state the connection existing between vanishing log-hybrid linear parameters and marginal independencies in the distribution of $Y_T | Y_U$.

Theorem 6.5   *In the setting of Theorem 6.3, for a pair of nonempty disjoint subsets $A, B \subseteq T$ the following conditions are equivalent:*

(i) $Y_A \perp\!\!\!\perp Y_B | Y_U$;
(ii) *it holds that $\varphi_{D \cup U'}^{(U,T)} = 0$ for every $D \subseteq A \cup B$ such that $D \cap A \neq \varnothing$ and $D \cap B \neq \varnothing$ and for every $U' \subseteq U$.*

Proof. We first show that for every nonempty $D \subseteq V$ the conditions (a) $\varphi_{D \cup U'} = 0$ for every $U' \subseteq U$, is equivalent to (b) $\gamma_D^{T|U'} = 0$ for every $U' \subseteq U$.

Point (a) implies (b) by (6.9) whereas (b) implies (a) because, by Möbius inversion, (6.9) gives $\varphi_{D \cup U'} = \sum_{U'' \subseteq U'} (-1)^{|U' \setminus U''|} \gamma_D^{T|U''}$ for every $U' \subseteq U$. Hence, condition (ii) is satisfied if and only if $\gamma_D^{T|U'} = 0$ for every $D \subseteq A \cup B$ such that $D \cap A \neq \emptyset$ and $D \cap B \neq \emptyset$ and for every $U' \subseteq U$, and this is equivalent to (i) by Theorem 5.2. $\square$

To see an example, assume that the distribution of Y is Markov with respect to the graph in Figure 6.4. Hence, condition (RG.b) implies $Y_e \perp\!\!\!\perp Y_h | Y_{\{a,b,c,d\}}$, and it follows from Theorem 6.5 that this independence relationship is satisfied if and only if $\varphi_{\{e,h\} \cup U'} = 0$ for every $U' \subseteq U$.

Finally, we consider independence relationships among context variables which, in the Markov property for regression graphs, are implied by missing undirected edges. Accordingly, the interest here lies in the conditional independence relationships in the distribution of Y_U and therefore in the log-linear parameter $\lambda^U = M_U^T \log \pi^U$. However, φ can be used in place of λ^U because it holds that $\lambda^U = (\varphi_D)_{D \subseteq U}$.

Proposition 6.6 *In the setting of Theorem 6.3, it holds that*

$$\varphi_D = \lambda_D^U \quad \text{for every} \quad U' \subseteq U,$$

where λ^U is the log-linear parameter of Y_U.

Proof. This follows from (4.2) and (6.5) by noticing that for every $D \subseteq U$

$$\pi_D^{(U,T)} = p(Y_D = 1, Y_{U \setminus D} = 0) = \pi_D^U$$

where π^U is the probability parameter of Y_U. $\square$

Hence, the connection between vanishing log-hybrid linear parameters and conditional independencies in the distribution of Y_U is as follows.

Theorem 6.7 *In the setting of Theorem 6.3, for a pair of nonempty disjoint subsets $A, B \subseteq U$ the following conditions are equivalent:*

1. $Y_A \perp\!\!\!\perp Y_B | Y_{U \setminus (A \cup B)}$;
2. *it holds that $\varphi_D^{(U,T)} = 0$ for every $D \subseteq U$ such that $D \cap A \neq \emptyset$ and $D \cap B \neq \emptyset$.*

Proof. This is an immediate consequence of Proposition 6.6 and Theorem 4.2. □

For example, in the regression graph of Figure 6.4 condition (RG.c) implies $Y_d \perp\!\!\!\perp Y_{\{a,c\}}|Y_b$, and it follows from Theorem 6.7 that this independence relationship is satisfied if and only if $\varphi_D = 0$ for every $D \subseteq U$ such that $d \in D$ and $D \cap \{a,c\} \neq \varnothing$.

The Non-binary Case

We now consider the case of categorical variables with an arbitrary number of levels. The material described here may be useful in the practical implementation of procedures based on the log-hybrid linear parameterization, and for this reason it is presented in a concise way as a list of technical results. More concretely, we provide a formal extension of the hybrid parameter and of the log-hybrid linear parameter of Y and, furthermore, of Theorems 6.3, 6.5 and 6.7.

Let Y be a vector of categorical variables with probability parameter π, and let $\mathcal{I}$ and $\mathcal{J}$ be the state space and restricted state space of Y, respectively. The hybrid parameter of Y relative to the partition (U, T) can be defined as the collection of vectors

$$\pi_j^{(U,T)} = (\pi_{j_D}^{(U,T)})_{D \subseteq V} \quad \text{for all } j \in \mathcal{J},$$

where

$$\pi_{j_D}^{(U,T)} = p(Y_D = j_D, Y_{U \setminus D} = 0).$$

The hybrid parameter of Y can be compactly written as $\pi^{(U,T)} = (\pi_{j_D}^{(U,T)})_{D \subseteq V, j_D \in \mathcal{J}_D}$, and it can be shown that there is a one-to-one correspondence between π and $\pi^{(U,T)}$. Hence, the log-hybrid linear parameter of Y is the collection of vectors

$$\varphi_j = (\varphi_{j_D})_{D \subseteq V} \quad \text{for all } j \in \mathcal{J},$$

where

$$\varphi_j = \mathbb{M}^T \log \pi_j^{(U,T)} \quad \text{and} \quad \varphi_{j_D} = (\varphi_j)_D,$$

and also in this case we use the compact form $\varphi = (\varphi_{j_D})_{D \subseteq V, j_D \in \mathcal{J}_D}$.

A formal extension of Theorem 6.3 to the general case of categorical variables with arbitrary number of levels is as follows.

Corollary 6.8 *Let Y_V be a vector of categorical variables with probability parameter $\pi > 0$, hybrid parameter $\pi^{(U,T)} = (\pi_{j_D}^{(U,T)})_{D \subseteq V, j_D \in \mathcal{J}_D}$ and log-hybrid linear parameter $\varphi = (\varphi_{j_D})_{D \subseteq V, j_D \in \mathcal{J}_D}$, where $V = U \cup T$ is a partition of V. Then, for a pair of nonempty subsets $A \subseteq T$ and $B \subseteq U$ the following conditions are equivalent:*

(i) $Y_A \perp\!\!\!\perp Y_B | Y_{U \setminus B}$;

(ii) *for every $D \subseteq A \cup U$ such that both $D \cap A \neq \varnothing$ and $D \cap B \neq \varnothing$ it holds that*

$$\varphi_{j_D} = 0,$$

for every $j \in \mathcal{J}$;

(iii) *for every $A' \subseteq A$, $B' \subseteq B$, and $C' \subseteq U \setminus B$, such that $A', B' \neq \varnothing$ it holds that*

$$\log \pi_{j_{A' \cup B' \cup C'}}^{(U,T)} - \log \pi_{j_{B' \cup C'}}^{(U,T)} - \log \pi_{j_{A' \cup C'}}^{(U,T)} + \log \pi_{j_{C'}}^{(U,T)} = 0, \quad (6.14)$$

for every $j \in \mathcal{J}$.

Proof. The equivalence of (i) and (iii) follows from Corollary 2.8 because the identity (6.14) is equal to $\log \mathrm{cpr}_{mc}(j_{A'}, A : j_{B'}, B \mid j_{C'}, C) = 0$ with $C = U \setminus B$. The equivalence of (ii) and (iii) holds for every $j \in \mathcal{J}$ and is that stated in Lemma 3.3 when $\theta = \varphi_j$ and $\omega = \log \pi_j^{(U,T)}$. □

The following corollary is a general version of Theorem 6.5.

Corollary 6.9 *In the setting of Corollary 6.8, for a pair of nonempty disjoint subsets $A, B \subseteq T$ the following conditions are equivalent:*

(i) $Y_A \perp\!\!\!\perp Y_B | Y_U$;

(ii) *it holds that $\varphi_{j_{D \cup U'}}^{(U,T)} = 0$ for every $D \subseteq A \cup B$ such that $D \cap A \neq \varnothing$ and $D \cap B \neq \varnothing$ and for every $U' \subseteq U$ and $j \in \mathcal{J}$.*

Proof. If, for $U' \in U$ and $j_{U'} \in \mathcal{J}_{U'}$, we denote by $\gamma^{T|j_{U'}} = (\gamma_{j_D}^{T|j_{U'}})_{D \subseteq T, j_D \in \mathcal{J}_D}$ the log-mean linear parameter of the conditional distribution of $Y_T | \{Y_{U'} = j_{U'}, Y_{U \setminus U'} = 0\}$, then following the same line used in the proof of Proposition 6.4 it can be shown that for every nonempty subset $D \subseteq T$ it holds that

$$\gamma_{j_D}^{T|_{U'}} = \sum_{U'' \subseteq U'} \varphi_{j_{D \cup U''}}.$$

Next, in a similar fashion used to prove Theorem 6.5 it can be shown that (ii) is equivalent to the statement

- it holds that $\gamma_{j_D}^{T|_{U'}} = 0$ for every $D \subseteq A \cup B$ such that $D \cap A \neq \emptyset$ and $D \cap B \neq \emptyset$ and for every $U' \subseteq U$ and $j \in \mathcal{J}$.

Hence, the result follows from Corollary 5.3.   □

Finally, we provide the extension of Theorem 6.7.

Corollary 6.10   *In the setting of Corollary 6.8, for a pair of nonempty disjoint subsets $A, B \subseteq U$ the following conditions are equivalent:*

(i) $Y_A \perp\!\!\!\perp Y_B | Y_{U \setminus (A \cup B)}$;

(ii) *it holds that $\varphi_{j_D}^{(U,T)} = 0$ for every $D \subseteq U$ such that $D \cap A \neq \emptyset$ and $D \cap B \neq \emptyset$ and for every $j \in \mathcal{J}$.*

Proof.   Let $\pi^U = (\pi_{j_D}^U)_{D \subseteq T j_D \in \mathcal{J}_D}$ be the probability parameter of Y_U. It follows from the fact that $\pi_{j_D}^{(U,T)} = \pi_{j_D}^U$ for every $D \subseteq U$ and $j \in \mathcal{J}$, that $\varphi_{j_D} = \lambda_{j_D}$ for every $D \subseteq U$ and $j \in \mathcal{J}$. Hence, the desired result can be obtained from the application of Corollary 4.3.   □

6.7 Log-hybrid Linear Graphical Models

Similarly to the other graphical models considered so far, a regression graph model for Y can be defined as the family of probability distributions that are Markov with respect to a given regression graph $G = (V, E)$. The theory presented in the previous section shows that a regression graph model corresponds to a linear subspace of the log-hybrid linear parameter space and, more specifically, it is characterized by the vanishing of certain φ-terms. Here, we look more closely at the connection between vanishing log-hybrid linear interactions and structure of the graph. Along the lines of Section 6.6, we start by considering the simplified setting of binary variables, and assume that the regression graph $G = (V, E)$ is *basic*, in the sense that it has two blocks, (U, T). The graph of Figure 6.4 is an example of a basic regression graph.

The definition of the Markov property for regression graphs in Section 6.4 is formulated by means of three statements, (RG.d),

(RG.b) and (RG.u), each concerning the independence relationships implied by the removal of one of the three types of edges. Likewise, there are three distinct rules for identifying the set of vanishing φ-terms from the structure of the graph. If $G = (V, E)$ is a basic regression graph with blocks (U, T) and Y_V is a binary random vector with log-hybrid linear parameter φ then the distribution of Y satisfies

(a) condition (RG.d) with respect to G if and only if
   [H.d] $\varphi_D = 0$ for every $D \subseteq V$ such that $D \cap T \neq \emptyset$ and $D \cap U \not\subseteq \mathrm{pa}(D \cap T)$;

(b) condition (RG.b) with respect to G if and only if
   [H.b] $\varphi_D = 0$ for every $D \subseteq V$ such that $D \cap T$ is disconnected in G_T;

(c) condition (RG.u) with respect to G if and only if
   [H.u] $\varphi_D = 0$ for every $D \subseteq U$ such that D is not complete in G_U.

The statement (b) follows from Theorem 6.5. It refers to the connected set Markov property for bidirected graphs, and when $U = \emptyset$ it is equivalent to Proposition 5.4. The statement (c) follows from Theorem 6.7 but, in fact, in a regression graph model the block U can be dealt with marginally as an undirected graph model for the distribution of Y_U and graph G_U. Recall that, as shown in Proposition 6.6, the φ-terms indexed by the subsets of U coincide with the log-linear interactions of Y_U. Statement (a) follows from Theorem 6.3, but its derivation is more involved than (b) and (c), and, for this reason, a formal proof is provided.

Proposition 6.11 *Let $G = (V, E)$ be a basic regression graph with blocks (U, T), and let φ be the log-hybrid linear parameter of a binary random vector Y_V. The distribution of Y_V satisfies condition (RG.d) with respect to G if and only if $\varphi_D = 0$ for every $D \subseteq V$ that such that both $D \cap T \neq \emptyset$ and $D \cap U \not\subseteq \mathrm{pa}(D \cap T)$.*

Proof. Assume that the distribution of Y_V satisfies condition (RG.d). If $D \subseteq V$ is such that $D \cap U \not\subseteq \mathrm{pa}(D \cap T)$ then $U \backslash \mathrm{pa}(D \cap T) \neq \emptyset$, and because it also holds that $D \cap T \neq \emptyset$ then (RG.d) implies that $Y_{D \cap T} \perp\!\!\!\perp Y_{U \backslash \mathrm{pa}(D \cap T)} | Y_{\mathrm{pa}(D \cap T)}$. Hence, it follows by point (ii) of Theorem 6.3 that $\varphi_D = 0$ because $D \cap T \neq \emptyset$ whereas $D \cap U \not\subseteq \mathrm{pa}(D \cap T)$ also implies that $D \cap \{U \backslash \mathrm{pa}(D \cap T)\} \neq \emptyset$. Conversely, assume that $\varphi_D = 0$ for every $D \subseteq V$ such that $D \cap T \neq \emptyset$ and $D \cap U \not\subseteq \mathrm{pa}(D \cap T)$. We have to show that $Y_A \perp\!\!\!\perp Y_B | Y_{U \backslash B}$ for all non-empty $A \subseteq T$ such that $B = U \backslash \mathrm{pa}(A)$ is also non-empty. Hence, by Theorem 6.3, we have to show that $\varphi_D = 0$ for every $D \subseteq A \cup U$ such that both $D \cap A \neq \emptyset$ and $D \cap B \neq \emptyset$. Now, $D \subseteq A \cup U$ implies $D \cap T = D \cap A \neq \emptyset$, while $D \cap B \neq \emptyset$ implies $D \cap U \not\subseteq \mathrm{pa}(A)$; because $\mathrm{pa}(D \cap A) \subseteq \mathrm{pa}(A)$ it follows that $D \cap U \not\subseteq \mathrm{pa}(D \cap A) = \mathrm{pa}(D \cap T)$. Therefore, we have $\varphi_D = 0$. $\qquad\square$

Clearly, the probability distribution of a binary random vector Y is Markov with respect to a basic regression graph G if and only if the zero structure of its log-hybrid linear parameter φ satisfies all the conditions [H.d], [H.b] and [H.u]. Conversely, in a basic regression graph model, the only φ-terms which are not constrained to vanish are those indexed by subsets $D \subseteq V$ such that either $D \subseteq U$ and D is complete or $D \cap T \neq \emptyset$ and both $D \cap T$ is connected and $D \cap U \subseteq \mathrm{pa}(D \cap T)$.

To fix ideas, assume that the distribution of Y is Markov with respect to the regression graph in Figure 6.4 and consider, for example, the set $D = \{b, d, f, h\}$. It follows from [H.d] that the corresponding φ-term vanishes, $\varphi_{\{b,d,f,h\}} = 0$. Indeed, every subset $D \subseteq V$, with $D \cap T \neq \emptyset$, identifies two subsets of U; namely, $\mathrm{pa}(D \cap T)$ and $D \cap U$. If $D \cap U \subseteq \mathrm{pa}(D \cap T)$ then [H.d] imposes no constraints on φ_D; otherwise φ_D vanishes. The set $D = \{b, d, f, h\}$ can be split into $D \cap T = \{f, h\}$ and $D \cap U = \{b, d\}$, and because $\{b, d\} \not\subseteq \mathrm{pa}(\{f, h\}) = \{d\}$ it follows that $\varphi_{\{b,d,f,h\}} = 0$. As a second example let $D = \{c, d, e, g, h\}$. In this case we find $D \cap T = \{e, g, h\}$ and $D \cap U = \{c, d\}$, so that $\{c, d\} \subseteq \mathrm{pa}(\{e, g, h\}) = \{c, d\}$, and therefore $\varphi_{\{c,d,e,g,h\}}$ is not restricted by [H.d]. The same interaction is not restricted by [H.b] because $\{c, d, e, g, h\} \cap T = \{e, g, h\}$ is a connected set. Because $\{c, d, e, g, h\} \not\subseteq U$ then $\varphi_{\{c,d,e,g,h\}}$ is not restricted by [H.u] either, and we can conclude that $\varphi_{\{c,d,e,g,h\}}$ is an unrestricted interaction of the model. Consider now the set $D = \{d, e, h\}$. In this case, $D \cap T = \{e, h\}$ and $D \cap U = \{d\}$, so that $\{d\} \subseteq \mathrm{pa}(\{e, h\}) = \{c, d\}$. Hence, $\varphi_{\{d,e,h\}}$ is not restricted by condition [H.d]; however it is restricted by [H.b] because $\{d, e, h\} \cap T = \{e, h\}$ is disconnected. More generally, because $\{e, h\}$ is disconnected in G_T it follows by [H.b] that $\varphi_{\{e,h\} \cup U'} = 0$ for every $U' \subseteq U$. Finally, the interaction $\varphi_{\{a,c,d\}}$ is not restricted by either [H.d] or [H.b] but it is constrained to vanish by [H.u] because $\{a, c, d\}$ is a subset of U and it is not complete.

We now turn to regression graphs with an arbitrary number of blocks. Let $G = (V, E)$ be an arbitrary regression graph, and let $(T_0, T_1, \ldots, T_k)$

be the block ordering of its vertices. The distribution of Y is characterized by the collection of conditional distributions $Y_{T_r}|Y_{\text{pr}(T_r)}$, for all $r = 1, \ldots, k$, together with the marginal distribution of Y_{T_0}. Accordingly, the φ-terms for the distribution of Y can be obtained by noticing that, for every $r = 1, \ldots, k$, the quantity $\varphi^{(\text{pr}(T_r), T_r)}$ parameterizes the distribution of $Y_{\text{pr}(T_r) \cup T_r}$ in such a way that $(\varphi_D^{(\text{pr}(T_r), T_r)})_{D \subseteq \text{pr}(T_r)} = \lambda^{\text{pr}(T_r)}$ parameterizes the distribution of $Y_{\text{pr}(T_r)}$, while $(\varphi_D^{(\text{pr}(T_r), T_r)})_{D \nsubseteq \text{pr}(T_r)}$ parameterizes the distribution of $Y_{T_r}|Y_{\text{pr}(T_r)}$. Hence, if we write $\varphi^{(T_r|\text{pr}(T_r))} = (\varphi_D^{(\text{pr}(T_r), T_r)})_{D \nsubseteq \text{pr}(T_r)}$ then the log-hybrid linear parameter of the distribution of Y_V are given by

$$\varphi^{(T_0, T_1)}, \varphi^{(T_2|\text{pr}(T_2))}, \ldots, \varphi^{(T_k|\text{pr}(T_k))}.$$

The vanishing φ-terms can be identified by applying the rules [H.d] and [H.b] iteratively for every $r = 1, \ldots k$ by setting $U = \text{pr}(T_r)$, $T = T_r$ and $V = \text{pr}(T_r) \cup T_r$. The rule [H.u] needs to be considered once with $U = T_0$. For example, in the regression graph of Figure 6.3, it holds that $\text{pr}(T_2) = T_0 \cup T_1$ and, for example, the [H.d] applied to the partition $(T_0 \cup T_1, T_2)$ implies that $\varphi_{\{l,h,c,b\}} = 0$. On the other hand, the term $\varphi_{\{l,h,c\}}$ is unconstrained because $\{l, h, c\} \cap \text{pr}(T_2) = \{h, c\} \subseteq \text{pa}(l)$ and, furthermore, $\{l, h, c\} \cap T_2 = \{l\}$ is connected.

A regression graph with singleton chain components is a DAG. In this case one can assign every vertex to a different block and then order the blocks according to any well-ordering of the vertices. In this way, the rules [H.b] and [H.u] become irrelevant, and it follows from [H.d] that the only unconstrained interactions are the terms $\varphi_{\{v\} \cup D}$ for all $v \in V$ and $D \subseteq \text{pa}(v)$.

Finally, conditions [H.d], [H.b] and [H.u] can be trivially extended to deal with non-binary variables by requiring $\varphi_{j_D} = 0$ for every $j_D \in \mathcal{J}_D$, in place of $\varphi_D = 0$.

6.8 Inference in Regression Graph Models

We conclude this chapter by dealing with inferential issues. Assume that $Y^{(1)} = y^{(1)}, \ldots, Y^{(n)} = y^{(n)}$ is a random sample of n independent and identically distributed observations from a distribution with probability parameter π. Then, the sampling distribution of counts follows a multinomial distribution with probability mass function given in

(4.12). The log-likelihood is therefore that given in (4.13), and it is equal to

$$l(\pi) = \sum_{i \in \mathcal{I}} n(Y = i) \log p(Y = i).$$

Let G be a regression graph. If the distribution of Y belongs to the regression graph model $M(G)$ then it follows from (6.2) that the log-likelihood function can be written as

$$l(\pi) = \sum_{K \in \mathcal{K}} l_K(\pi), \tag{6.15}$$

where $l_K(\pi)$, for $K \in \mathcal{K}$, are the *component log-likelihood functions*

$$l_K(\pi) = \sum_{i_K \in \mathcal{I}_K} \sum_{i_{\mathrm{pa}(K)} \in \mathcal{I}_{\mathrm{pa}(K)}} n(Y_{K \cup \mathrm{pa}(K)} = i_{K \cup \mathrm{pa}(K)})$$

$$\log p(Y_K = i_K | Y_{\mathrm{pa}(K)} = i_{\mathrm{pa}(K)}).$$

The decomposition of the likelihood in (6.15) can be exploited to simplify the computation of the ML estimates. Indeed, it can be shown (see, e.g. Drton, 2009) that $l(\pi)$ can be maximized in $M(G)$ by maximizing every $l_K(\pi)$ separately, and then by combining the optima according to (6.2).

The actual maximization of a component log-likelihood function is a problem in constrained optimization. For categorical data, a general iterative algorithm for constrained likelihood maximization was provided by Lang (1996). This algorithm can be applied to all models considered in this text; see also Evans and Forcina (2013) for a discussion on this issue. Furthermore, dedicated algorithms are available for the maximization of the component functions $l_K(\pi)$ in (6.15). In the case where the relevant chain component induces an undirected graph, the maximization of the component log-likelihood function can be carried out by applying the classical IPF algorithm. There also exists a dual of the IPF algorithm, named the *iterative conditional fitting* (ICF) algorithm, that can be applied to maximize the components $l_K(\pi)$ corresponding to the bidirected chain components. This algorithm was developed by Drton and Richardson (2008b) to fit discrete bidirected graph models and then extended by Drton (2008) to fit the bidirected components of regression graph models.

It is possible that the ML estimates of certain probabilities are simple empirical proportions and can therefore be computed directly. The following result is proved in Drton (2009, proposition 12).

Proposition 6.12 *If the chain component K of the regression graph G is a complete set and* $\mathrm{pa}(K) = \emptyset$, *then the ML estimator of the marginal probability* $p(Y_K = i_K)$ *is the empirical proportion*

$$\hat{p}(Y_K = i_K) = \frac{n(Y_K = i_K)}{n}.$$

Furthermore, if the chain component K is a singleton, then the ML estimator of the conditional probability $p(Y_K = i_K | Y_{\mathrm{pa}(K)} = i_{\mathrm{pa}(K)})$ *is the empirical proportion*

$$\hat{p}(Y_K = i_K | Y_{\mathrm{pa}(K)} = i_{\mathrm{pa}(K)}) = \frac{n(Y_K = i_K, Y_{\mathrm{pa}(K)} = i_{\mathrm{pa}(K)})}{n(Y_{\mathrm{pa}(K)} = i_{\mathrm{pa}(K)})}. \tag{6.16}$$

All the chain component of a DAG are singletons and therefore (6.16) implies that in DAG models the ML estimates are all empirical proportions. Indeed, equation (6.16) extends to regression graph models a classical result of DAG models; see, e.g., Lauritzen (1996, theorem 4.36). DAGs constitute a distinguished family of regression graphs and a comprehensive description of likelihood inference in these models can be found, for example, in Lauritzen (1996, section 4.5.1).

Finally, we remark that the standard asymptotic theory is valid for regression graph models so that, for instance, the asymptotic distribution of the ML estimator is normal and likelihood ratios are asymptotically chi-squared distributed. This follows from the fact that, as shown in Drton (2009), regression graphs yield models for categorical data that are curved exponential families.

Bibliography

Agresti, A. (2013). *Categorical Data Analysis*, 3rd edn, New York: John Wiley and Sons.

Ali, R. A., Richardson, T. S. & Spirtes, P. (2009). Markov equivalence for ancestral graphs. *The Annals of Statistics*, **37**(5B), 2808–37.

Anderson, T. W. (1969). Statistical inference for covariance matrices with linear structure. In *Multivariate Analysis, II: Proc. 2nd Int. Symp., Dayton, Ohio, 1968*. New York: Academic Press, pp. 55–66.

Anderson,T. W. (1973). Asymptotically efficient estimation of covariance matrices with linear structure. *The Annals of Statistics*, **1**(1), 135–41.

Andersson, S. A., Madigan, D., Perlman, M. D. (1997). A characterization of Markov equivalence classes for acyclic digraphs. *The Annals of Statistics*, **25** (2), 505–41.

Andersson, S. A., Madigan, D. & Perlman, M. D. (2001). Alternative Markov properties for chain graphs. *Scandinavian Journal of Statistics*, **28**(1), 33–85.

Asmussen, S. & Edwards, D. (1983). Collapsibility and response variables in contingency tables. *Biometrika*, **70**(3), 567–78.

Barber, D. (2012). *Bayesian Reasoning and Machine Learning*. Cambridge: Cambridge University Press.

Barndorff-Nielsen, O. (1978). *Information in Exponential Families and Conditioning*. New York: John Wiley and Sons.

Barndorff-Nielsen, O. (2014). *Information and Exponential Families in Statistical Theory*. Chichester: John Wiley and Sons.

Bartolucci, F., Colombi, R. & Forcina, A. (2007). An extended class of marginal link functions for modelling contingency tables by equality and inequality constraints. *Statistica Sinica*, **17**(2), 691–711.

Bergsma, W. P. & Rudas, T. (2002). Marginal models for categorical data. *The Annals of Statistics*, **30**(1), 140–59.

Birch, M. W. (1963). Maximum likelihood in three-way contingency tables. *Journal of the Royal Statistical Society. Series B (Statistical Methodology)*, **25**(1), 220–233.

Bishop, Y. M., Fienberg, S. E. & Holland, P. W. (1975). *Discrete Multivariate Analysis: Theory and Practice*. Cambridge, MA: MIT Press.

146

Bishop, Y. M., Fienberg, S. E. & Holland, P. W. (2007). *Discrete Multivariate Analysis: Theory and Practice*. New York: Springer-Verlag.

Boutilier, C., Friedman, N., Goldszmidt, M. & Koller, D. (1996). Context-specific independence in Bayesian networks.: *Proceedings of the Twelfth Annual Conference on Uncertainty in Artificial Intelligence (UAI-96)*. San Francisco, CA: Morgan Kaufmann, pp. 115–23.

Brown, L. D. (1986). *Fundamentals of Statistical Exponential Families with Applications in Statistical Decision Theory*. Lecture Notes-monograph series, vol. 9. Hayward, CA: Institute of Mathematical Statistics.

Chickering, D. M. (2002). Learning equivalence classes of Bayesian-network structures. *Journal of Machine Learning Research*, **2**, 445–98.

Christensen, R. (1997). *Log-linear Models and Logistic Regression*, 2nd edn, New York: Springer-Verlag.

Consonni, G. & Leucari, V. (2006). Reference priors for discrete graphical models. *Biometrika*, **93**(1), 23–40.

Coppen, A. (1966). The Marke–Nyman temperament scale: an English translation. *British Journal of Medical Psychology*, **39**(1), 55–9.

Corander, J. (2003). Labelled graphical models. *Scandinavian Journal of Statistics*, **30**(3), 493–508.

Cowell, R. G., Dawid, A. P., Lauritzen, S. L. & Spiegelhalter, D. J. (1999). *Probabilistic Networks and Expert Systems*. New York: Springer-Verlag.

Cox, D. R. & Wermuth, N. (1993). Linear dependencies represented by chain graphs. *Statistical Science*, **8**(3), 204–18.

Cox, D. R. & Wermuth, N. (1996). *Multivariate Dependencies: Models, Analysis, and Interpretation*. Boca Raton, FL: Chapman & Hall.

Darroch, J. N. & Ratcliff, D. (1972). Generalized iterative scaling for log-linear models. *The Annals of Mathematical Statistics*, **43**(5), 1470–80.

Darroch, J. N., Lauritzen, S. L. & Speed, T. P. (1980). Markov fields and log-linear interaction models for contingency tables. *The Annals of Statistics*, **8**(3), 522–39.

Davison, A. C. (2003). *Statistical Models*. Vol. 11. Cambridge: Cambridge University Press.

Dawid, A. P. (1979). Conditional independence in statistical theory. *Journal of the Royal Statistical Society: Series B (Statistical Methodology)*, **41**, 1–31.

Dawid, A. P. & Lauritzen, S. L. (1993). Hyper Markov laws in the statistical analysis of decomposable graphical models. *The Annals of Statistics*, **21**(3), 1272–317.

Deming, W. E. & Stephan, F. F. (1940). On a least squares adjustment of a sampled frequency table when the expected marginal totals are known. *The Annals of Mathematical Statistics*, **11**(4), 427–44.

Diestel, R. (1990). *Graph Decompositions: A Study in Infinite Graph Theory*. Oxford: Clarendon Press.

Drton, M. (2008). Iterative conditional fitting for discrete chain graph models. In P. Brito, ed., *COMPSTAT 2008 – Proceedings in Computational Statistics.* New York: Springer, pp. 93–104.

Drton, M. (2009). Discrete chain graph models. *Bernoulli*, **15**(3), 736–53.

Drton, M. & Maathuis, M. H. (2017). Structure learning in graphical modeling. *Annual Review of Statistics and Its Application*, **4**(1).

Drton, M. & Richardson, T. S. (2008a). Binary models for marginal independence. *Journal of the Royal Statistical Society: Series B (Statistical Methodology)*, **70**(2), 287–309.

Drton, M. & Richardson, T. S. (2008b). Graphical methods for efficient likelihood inference in Gaussian covariance models. *Journal of Machine Learning Research*, **9**, 893–914.

Drton, M., Lauritzen, S. L., Maathuis, M. & Wainwright, M. (2017). *Handbook of Graphical Models*. Boca Raton, FL: Chapman and Hall/CRC.

Edwards, D. (2000). *Introduction to Graphical Modelling*, 2nd edn, New York: Springer-Verlag.

Edwards, D. & Kreiner, S. (1983). The analysis of contingency tables by graphical models. *Biometrika*, **70**(3), 553–65.

Evans, R. J. (2016). Graphs for margins of Bayesian networks. *Scandinavian Journal of Statistics*, **43**(3), 625–48.

Evans, R. J. & Forcina, A. (2013). Two algorithms for fitting constrained marginal models. *Computational Statistics & Data analysis*, **66**, 1–7.

Evans, R. J. & Richardson, T. S. (2010). Maximum likelihood fitting of acyclic directed mixed graphs to binary data. In P. Grunwald & P. Spirtes, eds, *Proceedings of the 26th Conference on Uncertainty in Artificial Intelligence (UAI 2010)*. Corvallis, OR: AUAI Press, pp. 177–84.

Evans, R. J. & Richardson, T. S. (2013). Marginal log-linear parameters for graphical Markov models. *Journal of the Royal Statistical Society: Series B (Statistical Methodology)*, **75**(4), 743–68.

Evans, R. J. & Richardson, T. S. (2014). Markovian acyclic directed mixed graphs for discrete data. *The Annals of Statistics*, **42**(4), 1452–82.

Frydenberg, M. (1990). The chain graph Markov property. *Scandinavian Journal of Statistics*, **17**(4), 333–353.

Frydenberg, M. & Lauritzen, S. L. (1989). Decomposition of maximum likelihood in mixed graphical interaction models. *Biometrika*, **76**(3), 539–55.

Geiger, D. & Meek, C. (1998). Graphical models and exponential families. In *Proceedings of the Fourteenth Annual Conference on Uncertainty in Artificial Intelligence (UAI-98)*. San Francisco, CA: Morgan Kaufmann, pp. 156–65.

Geiger, D. & Pearl, J. (1988). On the logic of causal models. In *Uncertainty in Artificial Intelligence 4 Annual Conference on Uncertainty in Artificial Intelligence (UAI-88)*. Amsterdam: Elsevier Science, pp. 3–14.

Geiger, D. & Pearl, J. (1993). Logical and algorithmic properties of conditional independence and graphical models. *The Annals of Statistics*, **21**(4), 2001–21.

Graybill, F. A. (1983). *Matrices with Applications in Statistics*. Belmont, CA: Wadsworth.

Gutiérrez-Peña, E. & Smith, A. F. M. (1997). Exponential and Bayesian conjugate families: review and extensions. *Test*, **6**(1), 1–90.

Hall, P. (1934). A contribution to the theory of groups of prime-power order. *Proceedings of the London Mathematical Society*, **2**(1), 29–95.

Hammersley, J. M. & Clifford, P. (1971). Markov fields on finite graphs and lattices. Unpublished manuscript.

Højsgaard, S. (2004). Statistical inference in context specific interaction models for contingency tables. *Scandinavian Journal of Statistics*, **31**(1), 143–58.

Højsgaard, S., Edwards, D. & Lauritzen, S. L. (2012). *Graphical Models with R*. New York: Springer Science+Business Media.

Jeffreys, H. (1946). An invariant form for the prior probability in estimation problems. *Proceedings of the Royal Society of London A: Mathematical, Physical and Engineering Sciences*, **186**(1007), 453–61.

Jeffreys, H. (1961). *Theory of Probability*, 3rd edn. Oxford Classic Texts in the Physical Sciences. Oxford: Oxford University Press.

Jokinen, J. (2006). Fast estimation algorithm for likelihood-based analysis of repeated categorical responses. *Computational Statistics & Data Analysis*, **51**(3), 1509–22.

Kass, R. E. & Raftery, A. E. (1995). Bayes factors. *Journal of the American Statistical Association*, **90**(430), 773–95.

Kauermann, G. (1996). On a dualization of graphical Gaussian models. *Scandinavian Journal of Statistics*, **23**(1), 105–16.

Kauermann, G. (1997). A note on multivariate logistic models for contingency tables. *Australian Journal of Statistics*, **39**(3), 261–76.

Koski, T. and Noble, J. M. (2009). Graphical models and exponential families. In *Bayesian Networks: An Introduction*. Chichester: John Wiley and Sons, Ltd, chapter 8.

La Rocca, L. & Roverato, A. (2017). Discrete graphical models. In M. Drton, S. L. Lauritzen, M. Maathuis & M. Wainwright, eds, *Handbook of Graphical Models*. Handbooks of Modern Statistical Methods. Boca Raton, FL: Chapman and Hall/CRC.

Lang, J. B. (1996. Maximum likelihood methods for a generalized class of log-linear models. *The Annals of Statistics*, **24**(2), 726–52.

Lauritzen, S. L. (1996). *Graphical models*. Oxford: Clarendon Press.

Lauritzen, S. L. (2001). Causal inference from graphical models. In O.E. Barndorff-Nielsen, D. R. Cox & C. Klüppelberg, eds, *Complex Stochastic Systems*. London/Boca Raton: Chapman and Hall/CRC Press, pp. 63–107.

Lauritzen, S. L. & Richardson, T. S. (2002). Chain graph models and their causal interpretations. *Journal of the Royal Statistical Society: Series B (Statistical Methodology)*, **64**(3), 321–48.

Lauritzen, S. L. & Wermuth, N. (1989). Graphical models for associations between variables, some of which are qualitative and some quantitative. *The Annals of Statistics*, **17**(1), 31–57.

Lauritzen, S. L., Dawid, A. P., Larsen, B. N. & Leimer, H.-G. (1990). Independence properties of directed Markov fields. *Networks*, **20**(5), 491–505.

Lovász, L. (1993). *Combinatorial Problems and Exercises*, 2nd edn. Amsterdam: North-Holland.

Lupparelli, M. & Roverato, A. (2017). Log-mean linear regression models for binary responses with an application to multimorbidity. *Journal of the Royal Statistical Society: Series C (Applied Statistics)*, **66**(2), 227–252.

Lupparelli, M., Marchetti, G. M. & Bergsma, W. P. (2009). Parameterizations and fitting of bi-directed graph models to categorical data. *Scandinavian Journal of Statistics*, **36**(3), 559–76.

Lütkepol, H. (1996). *Handbook of Matrices*. Chichester: Wiley.

Madsen, M. (1976). Statistical analysis of multiple contingency tables. Two examples. *Scandinavian Journal of Statistics*, **3**(3), 97–106.

Marchetti, G. M. & Lupparelli, M. (2011). Chain graph models of multivariate regression type for categorical data. *Bernoulli*, **17**(3), 827–44.

Massam, H., Liu, J. & Dobra, A. (2009). A conjugate prior for discrete hierarchical log-linear models. *The Annals of Statistics*, **37**(6), 3431–67.

Meek, C. (1995). Causal inference and causal explanation with background knowledge. In *Proceedings of the Eleventh Annual Conference on Uncertainty in Artificial Intelligence (UAI-95)*. San Francisco, CA: Morgan Kaufmann, pp. 403–10.

Morris, C. N. (1982). Natural exponential families with quadratic variance functions. *The Annals of Statistics*, **10**(1), 65–80.

Morris, C. N. (1983). Natural exponential families with quadratic variance functions: statistical theory. *The Annals of Statistics*, **11**(2), 515–29.

Nyman, H., Pensar, J., Koski, T. & Corander, J. (2014). Stratified graphical models – context-specific independence in graphical models. *Bayesian Analysis*, **9**(4), 883–908.

Pearl, J. (1986). Fusion, propagation, and structuring in belief networks. *Artificial Intelligence*, **29**(3), 241–88.

Pearl, J. (1988). *Probabilistic Reasoning in Intelligent Systems*. San Mateo, CA: Morgan Kaufmann.

Pearl, J. (2009). *Causality*, 2nd edn, Cambridge: Cambridge University Press.

Pearl, J. & Paz, A. (1987). Graphoids: a graph-based logic for reasoning about relevancy relations. In B. D. Boulary, D. Hogg &L. Steel, eds, *Advances in Artificial Intelligence – II*. Amsterdam: North-Holland, pp. 357–63.

Pearl, J. & Verma, T. (1990). Equivalence and synthesis of causal models. In *Uncertainty in Artificial Intelligence 6 Annual Conference on Uncertainty in Artificial Intelligence (UAI-90)*. Amsterdam: Elsevier Science, pp. 255–68.

Piccioni, M. (2000). Independence structure of natural conjugate densities to exponential families and the Gibbs' sampler. *Scandinavian Journal of Statistics*, **27**(1), 111–27.

R Core Team. (2016). *R: A Language and Environment for Statistical Computing*. Vienna: Foundation for Statistical Computing.

Richardson, T. & Spirtes, P. (2002). Ancestral graph Markov models. *The Annals of Statistics*, **30**(4), 962–1030.

Richardson, T. S. (2003). Markov properties for acyclic directed mixed graphs. *Scandinavian Journal of Statistics*, **30**(1), 145–57.

Rota, G.-C. (1964). On the foundations of combinatorial theory I. Theory of Möbius functions. *Probability Theory and Related Fields*, **2**(4), 340–68.

Roverato, A. (2005). A unified approach to the characterization of equivalence classes of DAGs, chain graphs with no flags and chain graphs. *Scandinavian Journal of Statistics*, **32**(2), 295–312.

Roverato, A. (2015). Log-mean linear parameterization for discrete graphical models of marginal independence and the analysis of dichotomizations. *Scandinavian Journal of Statistics*, **42**(2), 627–48.

Roverato, A. & La Rocca, L. (2006). On block ordering of variables in graphical modelling. *Scandinavian Journal of Statistics*, **33**(1), 65–81.

Roverato, A. & Studený, M. (2006). A graphical representation of equivalence classes of AMP chain graphs. *Journal of Machine Learning Research*, **7**, 1045–78.

Roverato, A. & Whittaker, J. (1998). The Isserlis matrix and its application to non-decomposable graphical Gaussian models. *Biometrika*, **85**(3), 711–25.

Roverato, A., Lupparelli, M. & La Rocca, L. (2013). Log-mean linear models for binary data. *Biometrika*, **100**(2), 485–94.

Rudas, T., Bergsma, W. P. & Németh, R. (2010). Marginal log-linear parameterization of conditional independence models. *Biometrika*, **97**(4), 1006–12.

Sadeghi, K. & Lauritzen, S. L. (2014). Markov properties for mixed graphs. *Bernoulli*, **20**(2), 676–96.

Sadeghi, K. & Wermuth, N. (2016). Pairwise Markov properties for regression graphs. *Stat*, **5**, 286–94.

Speed, T. P. (1983). Cumulants and partition lattices. *Australian Journal of Statistics*, **25**(2), 378–88.

Spirtes, P., Glymour, C. & Scheines, R. (2000). *Causation, Prediction, and Search*, 2nd edn, Cambridge, MA: MIT Press.

Studený, M. (2005). *Probabilistic Conditional Independence Structures*. London: Springer-Verlag.

Tarjan, R. E. & Yannakakis, M. (1984). Simple linear-time algorithms to test chordality of graphs, test acyclicity of hypergraphs, and selectively reduce acyclic hypergraphs. *SIAM Journal on Computing*, **13**(3), 566–79.

Volf, M. & Studený, M. (1999). A graphical characterization of the largest chain graphs. *International Journal of Approximate Reasoning*, **20**(3), 209–36.

Weisner, L. (1935). Abstract theory of inversion of finite series. *Transactions of the American Mathematical Society*, **38**(3), 474–84.

Wermuth, N. (1976). Model search among multiplicative models. *Biometrics*, **32**(2), 253–63.

Wermuth, N. & Cox, D. R. (2015). Graphical Markov models: overview. In J. D. Wright, ed., *International Encyclopedia of the Social and Behavioral Sciences*, 2nd edn, vol. 10. Oxford: Elesevier, pp. 341–50.

Wermuth, N. & Lauritzen, S. L. (1990). On substantive research hypotheses, conditional independence graphs and graphical chain models. *Journal of the Royal Statistical Society: Series B (Statistical Methodology)*, **52**(1), 21–50.

Wermuth, N. & Sadeghi, K. (2012). Sequences of regressions and their independences. *TEST*, **21**(2), 215–52.

Whittaker, J. (1990). *Graphical Models in Applied Multivariate Statistics*. Chichester: Wiley.

Wright, S. (1921). Correlation and causation. *Journal of Agricultural Research*, **20**(7), 557–85.

Printed in the United States
By Bookmasters